# Optical Mineralogy

# Optical Mineralogy

Gunamay Bose

RANDOM PUBLICATIONS
NEW DELHI (INDIA)

**Optical Mineralogy**

---

ISBN 978-93-5111-421-5



Published in 2014 in India by

**RANDOM PUBLICATIONS**

4376-A/4B, Gali Murari Lal, Ansari Road
New Delhi-110 002
Phone : +91-11-43580356, +91-11-23289044
e-mail: randomexports@gmail.com, sales@randompublications.com, info@randompublications.com

Reprinted 2024

*Type Setting by:* Friends Media, Delhi-110089
*Digitally Printed at:* Replika Press Pvt. Ltd.

# Preface

Optical mineralogy is the study of the interaction of light with minerals, most commonly limited to visible light and usually further limited to the non-opaque minerals. Opaque minerals are more commonly studied in reflected light and that study is generally called ore microscopy - alluding to the fact many opaque minerals are also ore minerals. The most general application of optical mineralogy is to aid in the identification of minerals, either in rock thin sections or individual mineral grains. Another application occurs because the optical properties of minerals are related to the crystal chemistry of the mineral — for example, the mineral's chemical composition, crystal structure, order/disorder. Thus, relationships exist, and correlations are possible between them and some optical property. This often allows a simple optical measurement with the petrographic microscope that may yield important information about some crystal chemical aspect of the mineral under study.

Optical mineralogy is used to identify the mineralogical composition of geological materials in order to help reveal their origin and evolution. Some minerals are colourless and transparent, others are yellow or brown, green, blue, pink, etc. The same mineral may present a variety of colours, in the same or different rocks, and these colours may be arranged in zones parallel to the surfaces of the crystals. Thus tourmaline may be brown, yellow, pink, blue, green, violet, grey, or colourless, but every mineral has one or more characteristic, most common tints. The shapes of the crystals determine in a general way the outlines of the sections of them presented on the slides. If the mineral has one or more good cleavages they will be indicated by systems of cracks. The refractive index is also clearly shown by the appearance of the section, which are rough, with well-defined borders if they have a much stronger refraction than the medium in which they are mounted. Some minerals decompose readily and become turbid and semi-transparent; others remain always perfectly fresh and clear, others yield characteristic secondary products. The inclusions in the crystals are of great interest; one mineral may enclose another, or may contain spaces occupied by glass, by fluids or by gases.

It is an ideal book for advanced undergraduate and graduate courses in optical mineralogy, this accessible text is also an essential resource for petrology and petrography courses.

I thank all members of my team who have helped in the preparation of the book. My special thanks go to "Random Publication" who have published the book.

—*Gunamay Bose*

# Contents

# 1

# Introduction

## Mineralogy

Mineralogy is a subject of geology specializing in the scientific study of chemistry, crystal structure, and physical (including optical) properties of minerals. Specific studies within mineralogy include the processes of mineral origin and formation, classification of minerals, their geographical distribution, as well as their utilization.

### *History of Mineralogy*

Early writing on mineralogy, especially on gemstones, comes from ancient Babylonia, the ancient Greco-Roman world, ancient and medieval China, and Sanskrit texts from ancient India. Books on the subject included the Naturalis Historia of Pliny the Elder which not only described many different minerals but also explained many of their properties. The German Renaissance specialist Georgius Agricola wrote works such as *De re metallica* (*On Metals*, 1556) and *De Natura Fossilium* (*On the Nature of Rocks*, 1546) which began the scientific approach to the subject. Systematic scientific studies of minerals and rocks developed in post-Renaissance Europe. The modern study of mineralogy was founded on the principles of crystallography and microscopic study of rock sections with the invention of the microscope in the 17th century.

### *Europe and the Middle East*

The ancient Greek writers Aristotle (384–322 BC) and Theophrastus (370–285 BC) were the first in the Western tradition to write of minerals and their properties, as well as metaphysical explanations for them. The Greek philosopher Aristotle wrote his

*Meteorologica*, and in it theorized that all the known substances were composed of water, air, earth, and fire, with the properties of dryness, dampness, heat, and cold. The Greek philosopher and botanist Theophrastus wrote his *De Mineralibus*, which accepted Aristotle's view, and divided minerals into two categories: those affected by heat and those affected by dampness.

The metaphysical emanation and exhalation (*anathumiaseis*) theory of the Greek philosopher Aristotle included early speculation on earth sciences including mineralogy. According to his theory, while metals were supposed to be congealed by means of moist exhalation, dry gaseous exhalation (*pneumatodestera*) was the efficient material cause of minerals found in the Earth's soil. He postulated these ideas by using the examples of moisture on the surface of the earth (a moist vapour 'potentially like water'), while the other was from the earth itself, pertaining to the attributes of hot, dry, smoky, and highly combustible ('potentially like fire'). Aristotle's metaphysical theory from times of antiquity had wide-ranging influence on similar theory found in later medieval Europe, as the historian Berthelot notes:

*The theory of exhalations was the point of departure for later ideas on the generation of metals in the earth, which we meet with Proclus, and which reigned throughout the middle ages.*

Ancient Greek terminology of minerals has also stuck through the ages with widespread usage in modern times. For example, the Greek word asbestos (meaning 'inextinguishable', or 'unquenchable'), for the unusual mineral known today containing fibrous structure. The ancient historians Strabo (63 BC–19 AD) and Pliny the Elder (23–79 AD) both wrote of asbestos, its qualities, and its origins, with the Hellenistic belief that it was of a type of vegetable. Pliny the Elder listed it as a mineral common in India, while the historian Yu Huan (239–265 AD) of China listed this 'fireproof cloth' as a product of ancient Rome or Arabia (Chinese: Daqin). Although documentation of these minerals in ancient times does not fit the manner of modern scientific classification, there was nonetheless extensive written work on early mineralogy.

## Pliny the Elder

For example, Pliny devoted five entire volumes of his work Naturalis Historia (77 AD) to the classification of "earths, metals, stones, and gems". He not only describes many minerals not known to Theophrastus, but discusses their applications and properties. He is the first to correctly recognise the origin of amber for example, as the fossilized remnant of tree resin from the observation of insects trapped in some samples. He laid the basis of crystallography by discussing crystal habit, especially the octahedral shape of diamond. His discussion of mining methods is unrivalled in the ancient world, and includes, for example, an eye-witness account of gold mining in northern Spain, an account which is fully confirmed by modern research.

However, before the more definitive foundational works on mineralogy in the 16th century, the ancients recognized no more than roughly 350 minerals to list and describe.

### *Jabir and Avicenna*

With philosophers such as Proclus, the theory of Neoplatonism also spread to the Islamic world during the Middle Ages, providing a basis for metaphyiscal ideas on mineralogy in the medieval Middle East as well. The medieval Islamic scientists expanded upon this as well, including the Persian scientist Ibn Sina (980-1037 AD), also known as *Avicenna*, who rejected alchemy and the earlier notion of Greek metaphysics that metallic and other elements could be transformed into one another. However, what was largely accurate of the ancient Greek and medieval metaphysical ideas on mineralogy was the slow chemical change in composition of the Earth's crust. There was also the Islamic alchemist and scientist Jâbir ibn Hayyân (721-815 AD), who was the first to bring the experimental method into alchemy. Aided by Greek mathematics and Islamic mathematics, he discovered the syntheses for hydrochloric acid, nitric acid, distillation and crystallization (the latter two being essential for the understanding of modern mineralogy).

### *Georgius Agricola, 'Father of Mineralogy'*

In the early 16th century AD, the writings of the German scientist Georg Bauer, pen-name Georgius Agricola (1494-1555 AD), in his *Bermannus, sive de re metallica dialogus* (1530) is considered to be the official establishment of mineralogy in the modern sense of its study. He wrote the treatise while working as a town physician and making observations in Joachimsthal, which was then a centre for mining and metallurgic smelting industries. In 1544, he published his written work *De ortu et causis subterraneorum*, which is considered to be the foundational work of modern physical geology. In it (much like Ibn Sina) he heavily criticized the theories laid out by the ancient Greeks such as Aristotle. His work on mineralogy and metallurgy continued with the publication of *De veteribus et novis metallis* in

1546, and culminated in his best known works, the *De re metallica* of 1556. It was an impressive work outlining applications of mining, refining, and smelting metals, alongside discussions on geology of ore bodies, surveying, mine construction, and ventilation. He praises Pliny the Elder for his pioneering work Naturalis Historia and makes extensive references to his discussion of minerals and mining methods. For the next two centuries this written work remained the authoritative text on mining in Europe.

Agricola had many various theories on mineralogy based on empirical observation, including understanding of the concept of ore channels that were formed by the circulation of ground waters ('succi') in fissures subsequent to the deposition of the surrounding rocks. As will be noted below, the medieval Chinese previously had conceptions of this as well.

For his works, Agricola is posthumously known as the "Father of Mineralogy".

After the foundational work written by Agricola, it is widely agreed by the scientific community that the *Gemmarum et Lapidum Historia* of Anselmus de Boodt (1550–1632) of Bruges is the first definitive work of modern mineralogy. The German mining chemist J.F. Henckel wrote his *Flora Saturnisans* of 1760, which was the first treatise in Europe to deal with geobotanical minerals, although the Chinese had mentioned this in earlier treatises of 1421 and 1664. In addition, the Chinese writer Du Wan made clear references to weathering and erosion processes in his *Yun Lin Shi Pu* of 1133, long before Agricola's work of 1546.

### China and the Far East

In ancient China, the oldest literary listing of minerals dates back to at least the 4th century BC, with the *Ji Ni Zi* book listing twenty four of them. Chinese ideas of metaphysical mineralogy span back to at least the ancient Han Dynasty (202 BC–220 AD). From the 2nd century BC text of the *Huai Nan Zi*, the Chinese used ideological Taoist terms to describe meteorology, precipitation, different types of minerals, metallurgy, and alchemy. Although the understanding of these concepts in Han times was Taoist in nature, the theories proposed were similar to the Aristotelian theory of mineralogical exhalations (noted above). By 122 BC, the Chinese had thus formulated the theory for metamorphosis of minerals, although it is noted by historians such as Dubs that the tradition of alchemical-mineralogical Chinese doctrine stems back to the School of Naturalists headed by the philosopher Zou

Yan (305 BC–240 BC). Within the broad categories of rocks and stones (shi) and metals and alloys (jin), by Han times the Chinese had hundreds (if not thousands) of listed types of stones and minerals, along with theories for how they were formed.

In the 5th century AD, Prince Qian Ping Wang of the Liu Song Dynasty wrote in the encyclopedia *Tai-ping Yu Lan* (circa 444 AD, from the lost book *Dian Shu*, or *Management of all Techniques*):

*The most precious things in the world are stored in the innermost regions of all. For example, there is orpiment. After a thousand years it changes into realgar. After another thousand years the realgar becomes transformed into yellow gold.*

In ancient and medieval China, mineralogy became firmly tied to empirical observations in pharmaceutics and medicine. For example, the famous horologist and mechanical engineer Su Song (1020–1101 AD) of the Song Dynasty (960–1279 AD) wrote of mineralogy and pharmacology in his *Ben Cao Tu Jing* of 1070. In it he created a systematic approach to listing various different minerals and their use in medicinal concoctions, such as all the variously known forms of mica that could be used to cure various ills through digestion. Su Song also wrote of the subconchoidal fracture of native cinnabar, signs of ore beds, and provided description on crystal form. Similar to the ore channels formed by circulation of ground water mentioned above with the German scientist Agricola, Su Song made similar statements concerning copper carbonate, as did the earlier *Ri Hua Ben Cao* of 970 AD with copper sulphate.

The Yuan Dynasty scientist Zhang Si-xiao (died 1332 AD) provided a groundbreaking treatise on the conception of ore beds from the circulation of ground waters and rock fissures, two centuries before Georgius Agricola would come to similar conclusions. In his *Suo-Nan Wen Ji*, he applies this theory in describing the deposition of minerals by evaporation of (or precipitation from) ground waters in ore channels.

In addition to alchemical theory posed above, later Chinese writers such as the Ming Dynasty physician Li Shizhen (1518–1593 AD) wrote of mineralogy in similar terms of Aristotle's metaphysical theory, as the latter wrote in his pharmaceutical treatise *Bìncio Gangmu* (*Compendium of Materia Medica*, 1596). Another figure from the Ming era, the famous geographer Xu Xiake (1587–1641) wrote of mineral beds and mica schists in his treatise. However, while European literature on mineralogy became wide and varied, the writers of the Ming and Qing dynasties wrote little of the subject (even compared to Chinese of the earlier Song era). The only other works from these two eras worth mentioning were the *Shi Pin* (Hierarchy of Stones) of Yu Jun in 1617, the *Guai Shi Lu* (Strange Rocks) of Song Luo in 1665, and the *Guan Shi Lu* (On Looking at Stones) in 1668. However, one figure from the Song era that is worth mentioning above all is Shen Kuo.

### *Theories of Shen Kuo*

The medieval Chinese Song Dynasty statesman and scientist Shen Kuo (1031-1095 AD) wrote of his land formation theory involving concepts of mineralogy. In his *Meng Xi Bi Tan Dream Pool Essays,* 1088), Shen formulated a hypothesis for the process of land formation (geomorphology); based on his observation of marine fossil shells in a geological stratum in the Taihang Mountains hundreds of miles from the Pacific Ocean. He inferred that the land was formed by erosion of the mountains and by deposition of silt, and described soil erosion, sedimentation and uplift. In an earlier work of his (circa 1080), he wrote of a curious fossil of a sea-orientated creature found far inland. It is also of interest to note that the contemporary author of the *Xi Chi Cong Yu* attributed the idea of particular places under the sea where serpents and crabs were petrified to one Wang Jinchen. With Shen Kuo's writing of the discovery of fossils, he formulated a hypothesis for the shifting of geographical climates throughout time. This was due to hundreds of petrified bamboos found underground in the dry climate of northern China, once an enormous landslide upon the bank of a river revealed them. Shen theorized that in pre-historic times, the climate of Yanzhou must have been very rainy and humid like southern China, where bamboos are suitable to grow.

In a similar way, the historian Joseph Needham likened Shen's account with the Scottish scientist Roderick Murchison (1792–1871), who was inspired to become a geologist after observing a providential landslide. In addition, Shen's description of sedimentary deposition predated that of James Hutton, who wrote his groundbreaking work in 1802 (considered the foundation of modern geology). The influential philosopher Zhu Xi (1130–1200) wrote of this curious natural phenomena of fossils as well, and was known to have read the works of Shen Kuo. In comparison, the first mentioning of fossils found in the West was made nearly two centuries later with Louis IX of France in 1253 AD, who discovered fossils of marine animals (as recorded in Joinville's records of 1309 AD).

## *America*

Perhaps the most influential mineralogy text in the 19th and 20th centuries was the *Manual of Mineralogy* by James Dwight Dana, Harvard professor, first published in 1848. The fourth edition was entitled *Manual of Mineralogy and Lithology* (ed. 4, 1887). It became a standard college text, and has been continuously revised and updated by a succession of editors including W. E. Ford (13th-14th eds., 1912–1929), Cornelius S. Hurlbut (15th-21st eds., 1941–1999), and beginning with the 22nd by Cornelis Klein. The 23rd edition is now in print under the title *Manual of Mineral Science (Manual of Mineralogy)* (2007), revised by Cornelis Klein and Barbara Dutrow.

Equally influential was Dana's *System of Mineralogy*, first published in 1837, which has consistently been updated and revised. The 6th edition (1892) being edited by his son Edward Salisbury Dana. A 7th edition was published in 1944, and the 8th edition was published in 1997 under the title *Dana's New Mineralogy: The System of Mineralogy of James Dwight Dana and Edward Salisbury Dana*, edited by R. V. Gaines *et al.*

## Modern Mineralogy

Historically, mineralogy was heavily concerned with taxonomy of the rock-forming minerals; to this end, the International Mineralogical Association is an organization whose members represent mineralogists in individual countries. Its activities include managing the naming of minerals (via the Commission of New Minerals and Mineral Names), location of known minerals, etc. As of 2004 there are over 4,000 species of mineral recognized by the IMA. Of these, perhaps 150 can be called "common," another 50 are "occasional," and the rest are "rare" to "extremely rare."

More recently, driven by advances in experimental technique (such as neutron diffraction) and available computational power, the latter of which has enabled extremely accurate atomic-scale simulations of the behaviour of crystals, the science has branched out to consider more general problems in the fields of inorganic chemistry and solid-state physics. It, however, retains a focus on the crystal structures commonly encountered in rock-forming minerals (such as the perovskites, clay minerals and framework silicates). In particular, the field has made great advances in the understanding of the relationship between the atomic-scale structure of minerals and their function; in nature, prominent examples would be accurate measurement and prediction of the elastic properties of minerals, which has led to new insight into seismological behaviour of rocks and depth-related discontinuities in seismograms of the Earth's mantle. To this end, in their focus on the connection between atomic-scale phenomena and macroscopic properties, the mineral sciences (as they are now commonly known) display perhaps more of an overlap with materials science than any other discipline.

### *Physical Mineralogy*

Physical mineralogy is the specific focus on physical attributes of minerals. Description of physical attributes is the simplest way to identify, classify, and categorize minerals, and they include:

## Crystal Structure

In mineralogy and crystallography, crystal structure is a unique arrangement of atoms or molecules in a crystalline liquid or solid. A crystal structure is composed of a pattern, a set of atoms arranged in a particular way, and a lattice exhibiting long-range order and symmetry. Patterns are located upon the points of a lattice, which is an array of points repeating periodically in three dimensions. The points can be thought of as forming identical tiny boxes, called unit cells, that fill the space of the lattice. The lengths of the edges of a unit cell and the angles between them are called the *lattice parametres.* The symmetry properties of the crystal are embodied in its space group.

A crystal's structure and symmetry play a role in determining many of its physical properties, such as cleavage, electronic band structure, and optical transparency.

### *Unit Cell*

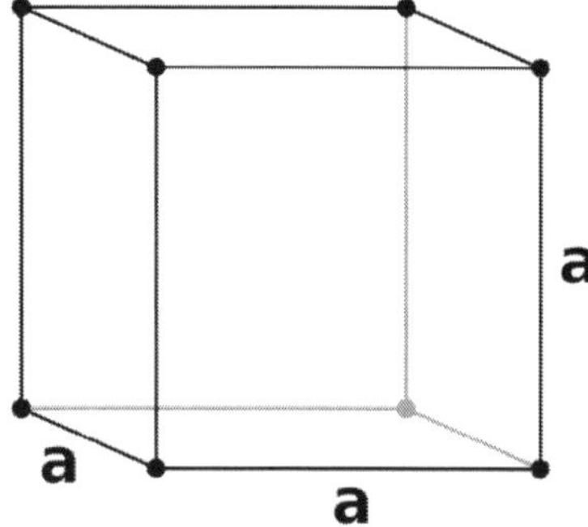


***Figure:*** *Simple cubic (P)*

The crystal structure of a material (the arrangement of atoms within a given type of crystal) can be described in terms of its unit cell. The unit cell is a small box containing one or more atoms arranged in 3-dimension. The unit cells stacked in three-dimensional space describe the bulk arrangement of atoms of the crystal. The unit cell is given by its lattice parametres, which are the length of the cell edges and the angles between them, while the positions of the atoms inside the unit cell are described by the set of atomic positions ($x_i$ , $y_i$ , $z_i$) measured from a lattice point.

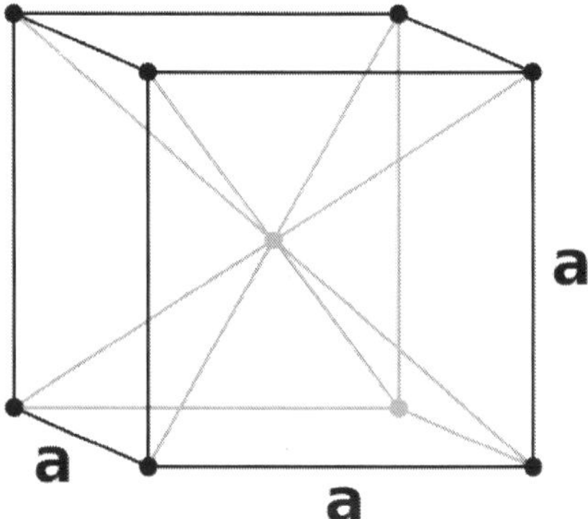


***Figure:*** *Body-centred cubic (I)*

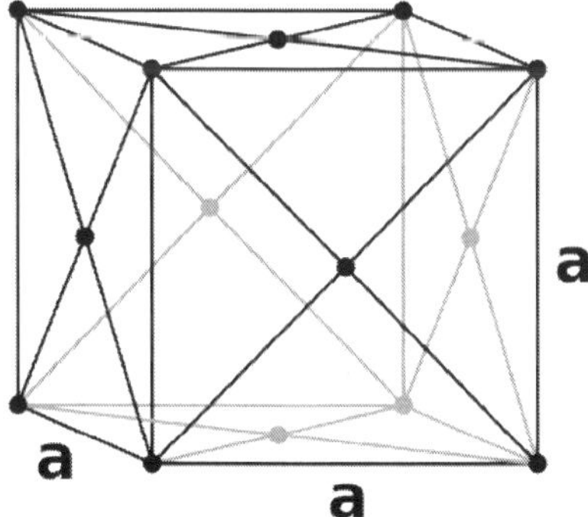


***Figure:*** *Face-centred cubic (F)*

Within the unit cell is the asymmetric unit, smallest unit the crystal can be divided into using the crystallographic symmetry operations of the space group. The asymmetric unit is also what is generally solved when solving a structure of a molecule or protein by X-ray crystallography.

### *Planes and Directions*

The crystallographic directions are geometric lines linking nodes (atoms, ions or molecules) of a crystal. Likewise, the crystallographic planes are geometric *planes* linking nodes. Some directions and planes have a higher density of nodes. These high density planes have an influence on the behaviour of the crystal as follows:

- Optical properties: Refractive index is directly related to density (or periodic density fluctuations).
- Adsorption and reactivity: Physical adsorption and chemical reactions occur at or near surface atoms or molecules. These phenomena are thus sensitive to the density of nodes.
- Surface tension: The condensation of a material means that the atoms, ions or molecules are more stable if they are surrounded by other similar species. The surface tension of an interface thus varies according to the density on the surface.

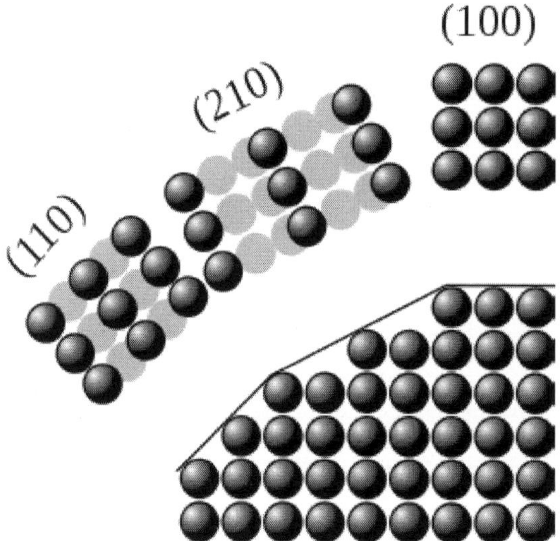


***Figure:*** *Dense crystallographic planes*

- Microstructural defects: Pores and crystallites tend to have straight grain boundaries following higher density planes.
- Cleavage: This typically occurs preferentially parallel to higher density planes.
- Plastic deformation: Dislocation glide occurs preferentially parallel to higher density planes. The perturbation carried by the dislocation (Burgers vector) is along a dense direction. The shift of one node in a more dense direction requires a lesser distortion of the crystal lattice.

Some directions and planes are defined by symmetry of the crystal system. In monoclinic, rombohedral, tetragonal, and trigonal/hexagonal systems there is one unique axis (sometimes called the principal axis) which has higher rotational symmetry than the other two axes. The basal plane is the plane perpendicular to the principal axis in these crystal systems. For triclinic, orthorhombic, and cubic crystal systems the axis designation is arbitrary and there is no principal axis.

### Cubic Structures

For the special case of simple cubic crystals, the lattice vectors are orthogonal and of equal length (usually denoted $a$); similarly for the reciprocal lattice. So, in this common case, the Miller indices ($\ell$mn) and [$\ell$mn] both simply denote normals/directions in Cartesian coordinates. For cubic crystals with lattice constant $a$, the spacing $d$ between adjacent ($\ell$mn) lattice planes is (from above):

$$d_{\ell mn} = \frac{a}{\sqrt{\ell^2 + m^2 + n^2}}$$

Because of the symmetry of cubic crystals, it is possible to change the place and sign of the integers and have equivalent directions and planes:

- Coordinates in *angle brackets* such as <100> denote a *family* of directions that are equivalent due to symmetry operations, such as [100], [010], [001] or the negative of any of those directions.
- Coordinates in *curly brackets* or *braces* such as {100} denote a family of plane normals that are equivalent due to symmetry operations, much the way angle brackets denote a family of directions.

For face-centred cubic (fcc) and body-centred cubic (bcc) lattices, the primitive lattice vectors are not orthogonal. However, in these cases the Miller indices are conventionally defined relative to the lattice vectors of the cubic supercell and hence are again simply the Cartesian directions.

### Classification

The defining property of a crystal is its inherent symmetry, by which we mean that under certain 'operations' the crystal remains unchanged. All crystals have translational symmetry in three directions, but some have other symmetry elements as well. For example, rotating the crystal 180° about a certain axis may result in an atomic configuration that is identical to the original configuration. The crystal is then said to have a twofold rotational symmetry about this axis. In addition to rotational symmetries like this, a crystal may have symmetries in the form of mirror planes and translational symmetries, and also the so-called "compound symmetries," which are a combination of translation and rotation/mirror symmetries. A full classification of a crystal is achieved when all of these inherent symmetries of the crystal are identified.

The simplest and most symmetric, the cubic (or isometric) system, has the symmetry of a cube, that is, it exhibits four threefold rotational axes oriented at 109.5° (the tetrahedral angle) with respect to each other. These threefold axes lie along the body diagonals of the cube. The other six lattice systems, are hexagonal, tetragonal, rhombohedral (often confused with the trigonal crystal system), orthorhombic, monoclinic and triclinic.

### Atomic Coordination

By considering the arrangement of atoms relative to each other, their coordination numbers (or number of nearest neighbours), interatomic distances, types of bonding, etc., it is possible to form a general view of the structures and alternative ways of visualizing them.

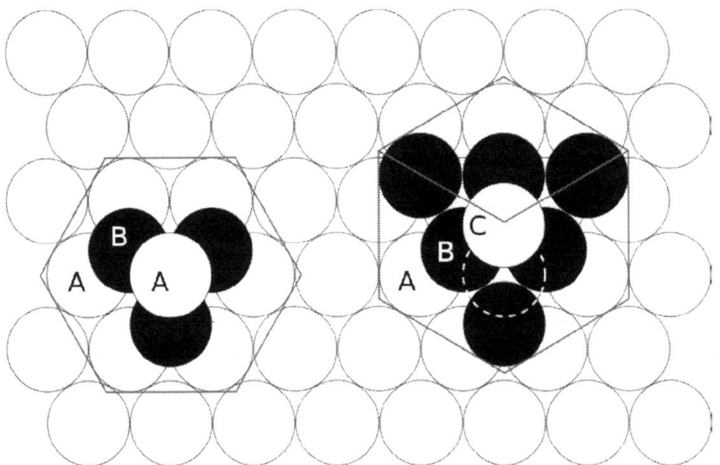


***Figure:*** *HCP lattice (left) and the fcc lattice (right)*

### Close Packing

The principles involved can be understood by considering the most efficient way of packing together equal-sized spheres and stacking close-packed atomic planes in three dimensions. For example, if plane A lies beneath plane B, there are two possible ways of placing an additional atom on top of layer B. If an additional layer was placed directly over plane A, this would give rise to the following series :

...ABABABAB....

This type of crystal structure is known as hexagonal close packing (hcp).

If however, all three planes are staggered relative to each other and it is not until the fourth layer is positioned directly over plane A that the sequence is repeated, then the following sequence arises:

...ABCABCABC...

This type of crystal structure is known as cubic close packing (ccp).

The unit cell of the ccp arrangement is the face-centred cubic (fcc) unit cell. This is not immediately obvious as the closely packed layers are parallel to the {111} planes of the fcc unit cell. There are four different orientations of the close-packed layers.

The packing efficiency could be worked out by calculating the total volume of the spheres and dividing that by the volume of the cell as follows:

$$\frac{4\times\frac{4}{3}\pi r^3}{16\sqrt{2}r^3}=\frac{\pi}{3\sqrt{2}}=0.7405$$

The 74% packing efficiency is the maximum density possible in unit cells constructed of spheres of only one size. Most crystalline forms of metallic elements are hcp, fcc, or bcc (body-centred cubic). The coordination number of hcp and fcc is 12 and its atomic packing factor (APF) is the number mentioned above, 0.74. The APF of bcc is 0.68 for comparison.

### *Bravais Lattices*

When the crystal systems are combined with the various possible lattice centring, we arrive at the Bravais lattices. They describe the geometric arrangement of the lattice points, and thereby the translational symmetry of the crystal. In three dimensions, there are 14 unique Bravais lattices that are distinct from one another in the translational symmetry they contain. All crystalline materials recognized until now (not including quasicrystals) fit in one of these arrangements. The fourteen three-dimensional lattices, classified by crystal system, are shown above. The Bravais lattices are sometimes referred to as *space lattices*.

The crystal structure consists of the same group of atoms, the *basis*, positioned around each and every lattice point. This group of atoms therefore repeats indefinitely in three dimensions according to the arrangement of one of the 14 Bravais lattices. The characteristic rotation and mirror symmetries of the group of atoms, or unit cell, is described by its crystallographic point group.

### *Point Groups*

The crystallographic point group or *crystal class* is the mathematical group comprising the symmetry operations that leave at least one point unmoved and that leave the appearance of the crystal structure unchanged. These symmetry operations include:

- *Reflection*, which reflects the structure across a *reflection plane*
- *Rotation*, which rotates the structure a specified portion of a circle about a *rotation axis*
- *Inversion*, which changes the sign of the coordinate of each point with respect to a *centre of symmetry* or *inversion point*
- *Improper rotation*, which consists of a rotation about an axis followed by an inversion.

Rotation axes (proper and improper), reflection planes, and centres of symmetry are collectively called *symmetry elements*. There are 32 possible crystal classes. Each one can be classified into one of the seven crystal systems.

### *Space Groups*

In addition to the operations of the point group, the space group of the crystal structure contains translational symmetry operations. These include:

- Pure *translations*, which move a point along a vector
- *Screw axes*, which rotate a point around an axis while translating parallel to the axis.
- *Glide planes*, which reflect a point through a plane while translating it parallel to the plane.

There are 230 distinct space groups.

### *Grain Boundaries*

Grain boundaries are interfaces where crystals of different orientations meet. A grain boundary is a single-phase interface, with crystals on each side of the boundary being identical except in orientation. The term "crystallite boundary" is sometimes, though rarely, used. Grain boundary areas contain those atoms that have been perturbed from their original lattice sites, dislocations, and impurities that have migrated to the lower energy grain boundary.

Treating a grain boundary geometrically as an interface of a single crystal cut into two parts, one of which is rotated, we see that there are five variables required to define a grain boundary. The first two numbers come from the unit vector that specifies a rotation axis. The third number designates the angle of rotation of the grain. The final two numbers specify the plane of the grain boundary (or a unit vector that is normal to this plane).

Grain boundaries disrupt the motion of dislocations through a material, so reducing crystallite size is a common way to improve strength, as described by the Hall–Petch relationship. Since grain boundaries are defects in the crystal structure they tend to decrease the electrical and thermal conductivity of the material. The high interfacial energy and relatively weak bonding in most grain boundaries often makes them preferred sites for the onset of corrosion and for the precipitation of new phases from the solid. They are also important to many of the mechanisms of creep. Grain boundaries are in general only a few nanometers wide. In common materials, crystallites are large enough that grain boundaries account for a small fraction of the material. However, very small grain sizes are achievable. In nanocrystalline solids, grain boundaries become a significant volume fraction of the material, with profound effects on such properties as diffusion and plasticity. In the limit of small crystallites, as the volume fraction of grain boundaries approaches 100%, the material ceases to have any crystalline character, and thus becomes an amorphous solid.

## Defects and Impurities

Real crystals feature defects or irregularities in the ideal arrangements described above and it is these defects that critically determine many of the electrical and mechanical properties of real materials. When one atom substitutes for one of the principal atomic components within the crystal structure, alteration in the electrical and thermal properties of the material may ensue. Impurities may also manifest as spin impurities in certain materials. Research on magnetic impurities demonstrates that substantial alteration of certain properties such as specific heat may be affected by small concentrations of an impurity, as for example impurities in semiconducting ferromagnetic alloys may lead to different properties as first predicted in the late 1960s. Dislocations in the crystal lattice allow shear at lower stress than that needed for a perfect crystal structure.

### *Prediction of Structure*

The difficulty of predicting stable crystal structures based on the knowledge of only the chemical composition has long been a stumbling block on the way to fully computational materials design. Now, with more powerful algorithms and high-performance computing, structures of medium complexity can be predicted using such approaches as evolutionary algorithms, random sampling, or metadynamics.

The crystal structures of simple ionic solids (e.g., NaCl or table salt) have long been rationalized in terms of Pauling's rules, first set out in 1929 by Linus Pauling, referred to by many since as the "father of the chemical bond". Pauling also considered the nature of the interatomic forces in metals, and concluded that about half of the five d-orbitals in the transition metals are involved in bonding, with the remaining nonbonding d-orbitals being responsible for the magnetic properties. He, therefore, was able to correlate the number of d-orbitals in bond formation with the bond length as well as many of the physical properties of the substance. He subsequently introduced the metallic orbital, an extra orbital necessary to permit uninhibited resonance of valence bonds among various electronic structures.

In the resonating valence bond theory, the factors that determine the choice of one from among alternative crystal structures of a metal or intermetallic compound revolve around the energy of resonance of bonds among interatomic positions. It is clear that some modes of resonance would make larger contributions (be more mechanically stable than others), and that in particular a simple ratio of number of bonds to number of positions would be exceptional. The resulting principle is that a special stability is associated with the simplest ratios or "bond numbers": 1/2, 1/3, 2/3, 1/4, 3/4, etc. The choice of structure and the value of the axial ratio (which determines the relative bond lengths) are thus a result of the effort of an atom to use its valency in the formation of stable bonds with simple fractional bond numbers.

After postulating a direct correlation between electron concentration and crystal structure in beta-phase alloys, Hume-Rothery analyzed the trends in melting points, compressibilities and bond lengths as a function of group number in the periodic table in order to establish a system of valencies of the transition elements in the metallic state. This treatment thus emphasized the increasing bond strength as a function of group number. The operation of directional forces were emphasized in one article on the relation between bond hybrids and the metallic structures. The resulting correlation between electronic and crystalline structures is summarized by a single parametre, the weight of the d-electrons per hybridized metallic orbital. The "d-weight" calculates out to 0.5, 0.7 and 0.9 for the fcc, hcp and bcc structures respectively. The relationship between d-electrons and crystal structure thus becomes apparent.

### *Polymorphism*

Polymorphism refers to the ability of a solid to exist in more than one crystalline form or structure. According to Gibbs' rules of phase equilibria, these unique crystalline phases will be dependent on intensive variables such as pressure and temperature. Polymorphism can potentially be found in many crystalline materials including polymers, minerals, and metals, and is related to allotropy, which refers to elemental solids. The complete morphology of a material is described by polymorphism and other variables such as crystal habit, amorphous fraction or crystallographic defects. Polymorphs have different stabilities and may spontaneously convert from a metastable form (or thermodynamically unstable form) to the stable form at a particular temperature. They also exhibit different melting points, solubilities, and X-ray diffraction patterns.

One good example of this is the quartz form of silicon dioxide, or $SiO_2$. In the vast majority of silicates, the Si atom shows tetrahedral coordination by 4 oxygens. All but one of the crystalline forms involve tetrahedral $SiO_4$ units linked together by shared vertices in different arrangements. In different minerals the tetrahedra show different degrees of networking and polymerization. For example, they occur singly, joined together in pairs, in larger finite clusters including rings, in chains, double chains, sheets, and three-dimensional frameworks. The minerals are classified into groups based on these structures. In each of its 7 thermodynamically stable crystalline forms or polymorphs of crystalline quartz, only 2 out of 4 of each the edges of the $SiO_4$ tetrahedra are shared with others, yielding the net chemical formula for silica: $SiO_2$.

Another example is elemental tin (Sn), which is malleable near ambient temperatures but is brittle when cooled. This change in mechanical properties due to existence of its two major allotropes, α- and β-tin. The two allotropes that are encountered at normal pressure and temperature, α-tin and β-tin, are more commonly known as *gray tin* and *white tin* respectively. Two more allotropes, γ and σ, exist at temperatures above 161 °C and pressures above several GPa. White tin is metallic, and is the stable crystalline form at or above room temperature. Below 13.2 °C, tin exists in the gray form, which has a diamond cubic crystal structure, similar to diamond, silicon or germanium. Gray tin has no metallic properties at all, is a dull-gray powdery material, and has few uses, other than a few specialized semiconductor applications. Although the α-β transformation temperature of tin is nominally 13.2 °C, impurities (e.g. Al, Zn, etc.) lower the transition temperature well below 0 °C, and upon addition of Sb or Bi the transformation may not occur at all.

### *Physical Properties*

Twenty of the 32 crystal classes are piezoelectric, and crystals belonging to one of these classes (point groups) display piezoelectricity. All piezoelectric classes lack a centre of symmetry. Any material develops a dielectric polarization when an electric field is applied, but a substance that has such a natural charge separation even in the absence of a field is called a polar material. Whether or not a material is polar is determined solely by its crystal structure. Only 10 of the 32 point groups are polar. All polar crystals are pyroelectric, so the 10 polar crystal classes are sometimes referred to as the pyroelectric classes.

There are a few crystal structures, notably the perovskite structure, which exhibit ferroelectric behaviour. This is analogous to ferromagnetism, in that, in the absence of an electric field during production, the ferroelectric crystal does not exhibit a polarization. Upon the application of an electric field of sufficient magnitude, the crystal becomes permanently polarized. This polarization can be reversed by a sufficiently large counter-charge, in the same way that a ferromagnet can be reversed. However, although they are called ferroelectrics, the effect is due to the crystal structure (not the presence of a ferrous metal).

## Crystal Habit

The crystal habit of a mineral describes its visible external shape. It can apply to an individual crystal or an assembly of crystals. In mineralogy, shape and size give rise to descriptive terms applied to the typical appearance, or habit of crystals. Each crystal can be described by how well it is formed, ranging from euhedral (perfect to near-perfect), to subhedral (moderately formed), and anhedral (poorly formed to no discernable habit seen).

The many terms used by mineralogists to describe crystal habits are useful in communicating what specimens of a particular mineral often look like. Recognizing numerous habits helps a mineralogist to identify a large number of minerals. Some habits are distinctive of certain minerals, although most minerals exhibit many differing habits (the development of a particular habit is determined by the details of the conditions during the mineral formation/crystal growth). Crystal habit may mislead the inexperienced as a mineral's internal crystal system can be hidden or disguised.

***Factors Influencing a Crystal's Habit Include:*** a combination of two or more crystal forms; trace impurities present during growth; crystal twinning and growth conditions (i.e., heat, pressure, space); and specific growth tendencies like growth striations. Minerals belonging to the same crystal system do not necessarily exhibit the same habit. Some habits of a mineral are unique to its variety and locality: For example, while most sapphires form elongate barrel-shaped crystals, those found in Montana form stout *tabular* crystals. Ordinarily, the latter habit is seen only in ruby. Sapphire and ruby are both varieties of the same mineral; corundum. Some minerals may replace other existing minerals while preserving the original's habit: this process is called pseudomorphous replacement. A classic example is tiger's eye quartz, crocidolite asbestos replaced by silica. While quartz typically forms *prismatic* (elongate, prism-like) crystals, in tiger's eye the original *fibrous* habit of crocidolite is preserved.

***The Names of Crystal Habits are Derived From:*** Predominant crystal faces (prism - prismatic, pyramid - pyramidal and pinacoid - platy). Crystal forms (cubic, octahedral, dodecahedral). Aggregation of crystals or aggregates (fibrous, botryoidal, radiating, massive). Crystal appearance (foliated/lamellar (layered), dendritic, bladed, acicular, lenticular, tabular (tablet shaped)).

## Crystal Twinning

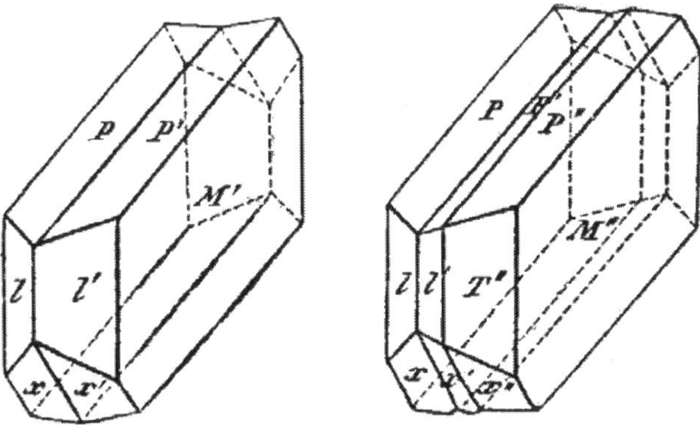

***Figure:*** *Diagram of twinned crystals of Albite. On the more perfect cleavage, which is parallel to the basal plane (P), is a system of fine striations, parallel to the second cleavage (M).*

Crystal twinning occurs when two separate crystals share some of the same crystal lattice points in a symmetrical manner. The result is an intergrowth of two separate crystals in a variety of specific configurations. A twin boundary or composition surface separates the two crystals. Crystallographers classify twinned crystals by a number of twin laws. These twin laws are specific to the crystal system. The type of twinning can be a diagnostic tool in mineral identification.

Twinning can often be a problem in X-ray crystallography, as a twinned crystal does not produce a simple diffraction pattern.

### Types of Twinning

Simple twinned crystals may be contact twins or penetration twins. Contact twins share a single composition surface often appearing as mirror images across the boundary. Plagioclase, quartz, gypsum, and spinel often exhibit contact twinning. Merohedral twinning occurs when the lattices of the contact twins superimpose in three dimensions, such as by relative rotation of one twin from the other. An example is metazeunerite. In penetration twins the individual crystals have the appearance of *passing through* each other in a symmetrical manner. Orthoclase, staurolite, pyrite, and fluorite often show penetration twinning.

If several twin crystal parts are aligned by the same twin law they are referred to as multiple or repeated twins. If these multiple twins are aligned in parallel they are called polysynthetic twins. When the multiple twins are not parallel they are cyclic twins. Albite, calcite, and pyrite often show polysynthetic twinning. Closely spaced polysynthetic twinning is often observed as striations or fine parallel lines on the crystal face. Rutile, aragonite, cerussite, and chrysoberyl often exhibit cyclic twinning, typically in a radiating pattern.

### Modes of Formation

There are three modes of formation of twinned crystals. Growth twins are the result of an interruption or change in the lattice during formation or growth due to a possible deformation from a larger substituting ion. Annealing or transformation twins are the result of a change in crystal system during cooling as one *form* becomes unstable and the crystal structure must re-organize or *transform* into another more stable form. Deformation or gliding twins are the result of stress on the crystal after the crystal has formed. If a FCC metal like aluminum experiences extreme stresses, it will experience twinning as seen in the case of explosions. Deformation twinning is a common result of regional metamorphism.

Crystals that grow adjacent to each other may be aligned to resemble twinning. This parallel growth simply reduces system energy and is not twinning.

### Twin Boundaries

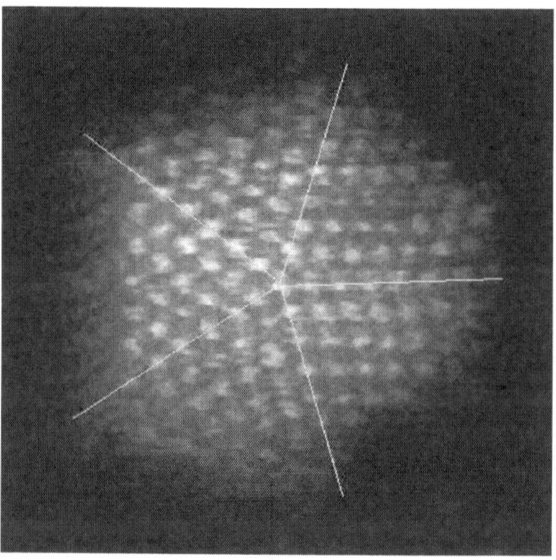

***Figure:*** *Fivefold twinning in a gold nanoparticle (electron microscope image).*

Twin boundaries occur when two crystals of the same type intergrow, so that only a slight misorientation exists between them. It is a highly symmetrical interface, often with one crystal the mirror image of the other; also, atoms are shared by the two crystals at regular intervals. This is also a much lower-energy interface than the grain boundaries that form when crystals of arbitrary orientation grow together.

Twin boundaries are partly responsible for shock hardening and for many of the changes that occur in cold work of metals with limited slip systems or at very low temperatures. They also occur due to martensitic transformations: the motion of twin boundaries is responsible for the pseudoelastic and shape-memory behaviour of nitinol, and their presence is partly responsible for the hardness due to quenching of steel. In certain types of high strength steels, very fine deformation twins act as primary obstacles against dislocation motion. These steels are referred to as 'TWIP' steels, where TWIP stands for TWinning Induced Plasticity.

### Deformation Twins

Of the three common crystalline structures BCC, FCC, and HCP, the HCP structure is the most likely to form deformation twins when strained, because they rarely have a sufficient number of slip systems for an arbitrary shape change. High strain rates, low Stacking-fault energy and low temperatures facilitate deformation twinning.

## Cleavage (Crystal)

Cleavage, in mineralogy, is the tendency of crystalline materials to split along definite crystallographic structural planes. These planes of relative weakness are a result of the regular locations of atoms and ions in the crystal, which create smooth repeating surfaces that are visible both in the microscope and to the naked eye.

### *Types of Cleavage*

Cleavage forms parallel to crystallographic planes:

- Basal or pinacoidal cleavage occurs parallel to the base of a crystal. This orientation is given by the {001} plane in the crystal lattice, and is the same as the {0001} plane in Bravais-Miller indices, which are often used for rhombohedral and hexagonal crystals. Basal cleavage is exhibited by the mica group and by graphite.
- Cubic cleavage occurs on the {001} planes, parallel to the faces of a cube for a crystal with cubic symmetry. This is the source of the cubic shape seen in crystals of ground table salt, the mineral halite. The mineral galena also typically exhibits perfect cubic cleavage.
- Octahedral cleavage occurs on the {111} crystal planes, forming octahedra shapes for a crystal with cubic symmetry. Diamond and fluorite exhibit perfect octahedral cleavage. Octahedral cleavage is seen in common semiconductors. For lower-symmetry crystals, there will be a smaller number of {111} planes.
- Dodecahedral cleavage occurs on the {110} crystal planes forming dodecahedra for a crystal with cubic symmetry. For lower-symmetry crystals, there will be a smaller number of {110} planes.
- Rhombohedral cleavage occur parallel to the {1011} faces of a rhombohedron. Calcite and other carbonate minerals exhibit perfect rhombohedral cleavage.
- Prismatic cleavage is cleavage parallel to a vertical prism {110}. Cerussite, tremolite and spodumene exhibit prismatic cleavage.

### Parting

Crystal parting occurs when minerals break along planes of structural weakness due to external stress or along twin composition planes. Parting breaks are very similar in appearance to cleavage, but only occur due to stress. Examples include magnetite which shows octahedral parting, the rhombohedral parting of corundum and basal parting in pyroxenes.

### Uses

Cleavage is a physical property traditionally used in mineral identification, both in hand specimen and microscopic examination of rock and mineral studies. As an example, the angles between the prismatic cleavage planes for the pyroxenes (88–92°) and the amphiboles (56–124°) are diagnostic.

Crystal cleavage is of technical importance in the electronics industry and in the cutting of gemstones.

Precious stones are generally cleaved by impact as in diamond cutting.

Synthetic single crystals of semiconductor materials are generally sold as thin wafers which are much easier to cleave. Simply pressing a silicon wafer against a soft surface and scratching its edge with a diamond scribe is usually enough to cause cleavage; however, when dicing a wafer to form chips, a procedure of scoring and breaking is often followed for greater control. Elemental semiconductors (Si, Ge, and diamond) are diamond cubic, a space group for which octahedral cleavage is observed. This means that some orientations of wafer allow near-perfect rectangles to be cleaved. Most other commercial semiconductors (GaAs, InSb, etc.) can be made in the related zinc blende structure, with similar cleavage planes.

## Lustre (Mineralogy)

Lustre (or luster) is the way light interacts with the surface of a crystal, rock, or mineral. The word traces its origins back to the latin *lux*, meaning “light”, and generally implies radiance, gloss, or brilliance.

A range of terms are used to describe lustre, such as *earthy*, *metallic*, *greasy*, and *silky*. Similarly, the term *vitreous* (derived from the Latin for glass, *vitrum*) refers to a glassy lustre. A list of these terms is given below.

Lustre varies over a wide continuum, and so there are no rigid boundaries between the different types of lustre. (For this reason, different sources can often describe the same mineral differently. This ambiguity is further complicated by lustre's ability to vary widely within a particular mineral species.) The terms are frequently combined to describe intermediate types of lustre (for example, a "vitreous greasy" lustre).

Some minerals exhibit unusual optical phenomena, such as asterism (the display of a star-shaped luminous area) or chatoyancy (the display of luminous bands, which appear to move as the specimen is rotated). A list of such phenomena is given below.

### Common Terms

***Adamantine Luster:*** Adamantine minerals possess a superlative lustre, which is most notably seen in diamond. Such minerals are transparent or translucent, and have a high refractive index (of 1.9 or more). Minerals with a true adamantine lustre are uncommon, with examples being cerussite and zircon.

Minerals with a lesser (but still relatively high) degree of lustre are referred to as subadamantine, with some examples being garnet and corundum.

### Dull Luster

Dull (or earthy) minerals exhibit little to no lustre, due to coarse granulations which scatter light in all directions, approximating a Lambertian reflector. An example is kaolinite. A distinction is sometimes drawn between dull minerals and earthy minerals, with the latter being coarser, and having even less lustre.

### Greasy Luster

Greasy minerals resemble fat or grease. A greasy lustre often occurs in minerals containing a great abundance of microscopic inclusions, with examples including opal and cordierite. Many minerals with a greasy lustre also feel greasy to the touch.

### Metallic Luster

Metallic (or splendant) minerals have the lustre of polished metal, and with ideal surfaces will work as a reflective surface. Examples include galena, pyrite and magnetite.

### Pearly Luster

Pearly minerals consist of thin transparent co-planar sheets. Light reflecting from these layers give them a lustre reminiscent of pearls. Such minerals possess perfect cleavage, with examples including muscovite and stilbite.

### *Resinous Luster*

Resinous minerals have the appearance of resin, chewing gum or (smooth surfaced) plastic. A principal example is amber, which is a form of fossilized resin.

### *Silky Lustre*

Silky minerals have a parallel arrangement of extremely fine fibres, giving them a lustre reminiscent of silk. Examples include asbestos, ulexite and the satin spar variety of gypsum. A fibrous lustre is similar, but has a coarser texture.

### *Submetallic Lustre*

Submetallic minerals have similar lustre to metal, but are duller and less reflective. A submetallic lustre often occurs in near-opaque minerals with very high refractive indices, such as sphalerite, cinnabar and cuprite.

### *Vitreous Lustre*

Vitreous minerals have the lustre of glass. (The term is derived from the Latin for glass, *vitrum*.) This type of lustre is one of the most commonly seen, and occurs in transparent or translucent minerals with relatively low refractive indices. Common examples include calcite, quartz, topaz, beryl, tourmaline and fluorite, among others.

### *Waxy Luster*

Waxy minerals have a lustre resembling wax. Examples include jade and chalcedony.

## Optical Phenomena

***Asterism:*** Asterism is the display of a star-shaped luminous area. It is seen in some sapphires and rubies, where it is caused by impurities of rutile. It can also occur in garnet, diopside and spinel.

***Aventurescence:*** Aventurescence (or aventurization) is a reflectance effect like that of glitter. It arises from minute, preferentially oriented mineral platelets within the material. These platelets are so numerous that they also influence the material's body colour. In aventurine quartz, chrome-bearing fuchsite makes for a green stone and various iron oxides make for a red stone.

### *Chatoyancy*

Chatoyant minerals display luminous bands, which appear to move as the specimen is rotated. Such minerals are composed of parallel fibres (or contain fibrous voids or inclusions), which reflect light into a direction perpendicular to their orientation, thus forming narrow bands of light. The most famous examples are tiger's eye and cymophane, but the effect may also occur in other minerals such as aquamarine, moonstone and tourmaline.

### *Colour Change*

Colour change is most commonly found in Alexandrite, a variety of chrysoberyl gemstones. Other gems also occur in colour-change varieties, including (but not limited to) sapphire, garnet, spinel. Alexandrite displays a colour change dependent upon light, along with strong pleochroism. The gem results from small scale replacement of aluminium by chromium oxide, which is responsible for alexandrite's characteristic green to red colour change. Alexandrite from the Ural Mountains in Russia is green by daylight and red by incandescent light. Other varieties of alexandrite may be yellowish or pink in daylight and a columbine or raspberry red by incandescent light. The optimum or "ideal" colour change would be fine emerald green to fine purplish red, but this is exceedingly rare.

### *Schiller*

Schiller, from German for "twinkle", is the metallic iridescence originating from below the surface of a stone, that occurs when light is reflected between layers of minerals. It is seen in moonstone and labradorite and is very similar to adularescence and aventurescence.

## Transparency and Translucency

In the field of optics, transparency (also called pellucidity or diaphaneity) is the physical property of allowing light to pass through the material without being scattered. On a macroscopic scale (one where the dimensions investigated are much, much larger than the wavelength of the photons in question), the photons can be said to follow Snell's Law. Translucency (also called translucence or translucidity) is a super-set of transparency: it allows light to pass through, but does not necessarily (again, on the macroscopic scale) follow Snell's law; the photons can be scattered at either of the two interfaces where there is a change in index of refraction, or internally. In other words, a translucent medium allows the transport of light while a transparent medium not only allows the transport of light but allows for image formation. The opposite property of translucency is opacity. Transparent materials appear clear, with the overall appearance of one colour, or any combination leading up to a brilliant spectrum of every colour. When light encounters a material, it can interact with it in several different ways. These interactions depend on the wavelength of the light and the nature of the material. Photons interact with an object by some combination of reflection, absorption and transmission. Some materials, such as plate glass and clean water, allow much of the light that falls on them to be transmitted, with little being reflected; such materials are called optically transparent. Many liquids and aqueous solutions are highly transparent. Absence of structural defects (voids, cracks, etc.) and molecular structure of most liquids are mostly responsible for excellent optical transmission.

Materials which do not allow the transmission of light are called opaque. Many such substances have a chemical composition which includes what are referred to as absorption centres. Many substances are selective in their absorption of white light frequencies. They absorb certain portions of the visible spectrum, while reflecting others. The frequencies of the spectrum which are not absorbed are either reflected back or transmitted for our physical observation. This is what gives rise to colour. The attenuation of light of all frequencies and wavelengths is due to the combined mechanisms of absorption and scattering.

Transparency offers the possibility of providing almost perfect camouflage for animals able to achieve it. This is easier in dimly-lit or turbid seawater than in good illumination. Many marine animals such as jellyfish are highly transparent.

With regard to the absorption of light, primary material considerations include:

- At the electronic level, absorption in the ultraviolet and visible (UV-Vis) portions of the spectrum depends on whether the electron orbitals are spaced (or "quantized") such that they can absorb a quantum of light (or photon) of a specific frequency, and does not violate selection rules. For example, in most glasses, electrons have no available energy levels above them in range of that associated with visible light, or if they do, they violate selection rules, meaning there is no appreciable absorption in pure (undoped) glasses, making them ideal transparent materials for windows in buildings.
- At the atomic or molecular level, physical absorption in the infrared portion of the spectrum depends on the frequencies of atomic or molecular vibrations or chemical bonds, and on

selection rules. Nitrogen and oxygen are not greenhouse gases because the absorption is forbidden by the lack of a molecular dipole moment.

With regard to the scattering of light, the most critical factor is the length scale of any or all of these structural features relative to the wavelength of the light being scattered. Primary material considerations include:

- Crystalline structure: whether or not the atoms or molecules exhibit the 'long-range order' evidenced in crystalline solids.
- Glassy structure: scattering centres include fluctuations in density or composition.
- Microstructure: scattering centres include internal surfaces such as grain boundaries, crystallographic defects and microscopic pores.
- Organic materials: scattering centres include fibre and cell structures and boundaries.

### Light Scattering in Solids

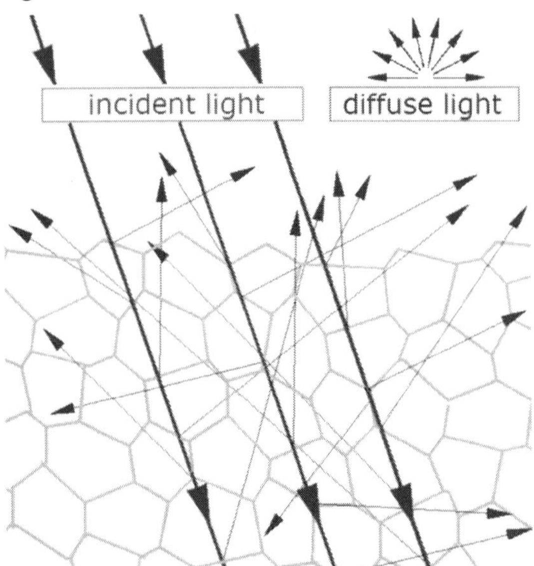


***Figure:*** *General mechanism of diffuse reflection*

Diffuse reflection - Generally, when light strikes the surface of a (non-metallic and non-glassy) solid material, it bounces off in all directions due to multiple reflections by the microscopic irregularities *inside* the material (e.g., the grain boundaries of a polycrystalline material, or the cell or fibre boundaries of an organic material), and by its surface, if it is rough. Diffuse reflection is typically characterized by omni-directional reflection angles. Most of the objects visible to the naked eye are identified via diffuse reflection. Another term commonly used for this type of reflection is "light scattering". Light scattering from the surfaces of objects is our primary mechanism of physical observation.

Light scattering in liquids and solids depends on the wavelength of the light being scattered. Limits to spatial scales of visibility (using white light) therefore arise, depending on the frequency of the light wave and the physical dimension (or spatial scale) of the scattering centre. Visible light has a wavelength scale on the order of a half a micrometre (one millionth of a metre). Scattering centres (or particles) as small as one micrometre have been observed directly in the light microscope (e.g., Brownian motion).

### *Applications*

Optical transparency in polycrystalline materials is limited by the amount of light which is scattered by their microstructural features. Light scattering depends on the wavelength of the light. Limits to spatial scales of visibility (using white light) therefore arise, depending on the frequency of the light wave and the physical dimension of the scattering centre. For example, since visible light has a wavelength scale on the order of a micrometre, scattering centres will have dimensions on a similar spatial scale. Primary scattering centres in polycrystalline materials include microstructural defects such as pores and grain boundaries. In addition to pores, most of the interfaces in a typical metal or ceramic object are in the form of grain boundaries which separate tiny regions of crystalline order. When the size of the scattering centre (or grain boundary) is reduced below the size of the wavelength of the light being scattered, the scattering no longer occurs to any significant extent.

In the formation of polycrystalline materials (metals and ceramics) the size of the crystalline grains is determined largely by the size of the crystalline particles present in the raw material during formation (or pressing) of the object. Moreover, the size of the grain boundaries scales directly with particle size. Thus a reduction of the original particle size well below the wavelength of visible light (about 1/15 of the light wavelength or roughly 600/15 = 40 nm) eliminates much of light scattering, resulting in a translucent or even transparent material.

Computer modelling of light transmission through translucent ceramic alumina has shown that microscopic pores trapped near grain boundaries act as primary scattering centres. The volume fraction of porosity had to be reduced below 1% for high-quality optical transmission (99.99 percent of theoretical density). This goal has been readily accomplished and amply demonstrated in laboratories and research facilities worldwide using the emerging chemical processing methods encompassed by the methods of sol-gel chemistry and nanotechnology.

***Figure:*** *Translucency of a material being used to highlight the structure of a photographic subject*

Transparent ceramics have created interest in their applications for high energy lasers, transparent armour windows, nose cones for heat seeking missiles, radiation detectors for non-destructive testing, high energy physics, space exploration, security and medical imaging applications.

The development of transparent panel products will have other potential advanced applications including high strength, impact-resistant materials that can be used for domestic windows and skylights. Perhaps more important is that walls and other applications will have improved overall strength, especially for high-shear conditions found in high seismic and wind exposures. If the expected improvements in mechanical properties bear out, the traditional limits seen on glazing areas in today's building codes could quickly become outdated if the window area actually contributes to the shear resistance of the wall.

Currently available infrared transparent materials typically exhibit a trade-off between optical performance, mechanical strength and price. For example, sapphire (crystalline alumina) is very strong, but it is expensive and lacks full transparency throughout the 3–5 micrometre mid-infrared range. Yttria is fully transparent from 3–5 micrometres, but lacks sufficient strength, hardness, and thermal shock resistance for high-performance aerospace applications. Not surprisingly, a combination of these two materials in the form of the yttrium aluminium garnet (YAG) is one of the top performers in the field.

## Absorption of Light in Solids

When light strikes an object, it usually has not just a single frequency (or wavelength) but many. Objects have a tendency to selectively absorb, reflect or transmit light of certain frequencies. That is, one object might reflect green light while absorbing all other frequencies of visible light. Another object might selectively transmit blue light while absorbing all other frequencies of visible light. The manner in which visible light interacts with an object is dependent upon the frequency of the light, the nature of the atoms in the object, and often the nature of the electrons in the atoms of the object.

Some materials allow much of the light that falls on them to be transmitted through the material without being reflected. Materials that allow the transmission of light waves through them are called optically transparent. Chemically pure (undoped) window glass and clean river or spring water are prime examples of this.

Materials which do not allow the transmission of any light wave frequencies are called opaque. Such substances may have a chemical composition which includes what are referred to as absorption centres. Most materials are composed of materials which are selective in their absorption of light frequencies. Thus they absorb only certain portions of the visible spectrum. The frequencies of the spectrum which are not absorbed are either reflected back or transmitted for our physical observation. In the visible portion of the spectrum, this is what gives rise to colour.

Colour centres are largely responsible for the appearance of specific wavelengths of visible light all around us. Moving from longer (0.7 micrometre) to shorter (0.4 micrometre) wavelengths: red, orange, yellow, green and blue (ROYGB) can all be identified by our senses in the appearance of colour by the selective absorption of specific light wave frequencies (or wavelengths). Mechanisms of selective light wave absorption include:

- Electronic: Transitions in electron energy levels within the atom (e.g., pigments). These transitions are typically in the ultraviolet (UV) and/or visible portions of the spectrum.
- Vibrational: Resonance in atomic/molecular vibrational modes. These transitions are typically in the infrared portion of the spectrum.

### *UV-Vis: Electronic Transitions*

In electronic absorption, the frequency of the incoming light wave is at or near the energy levels of the electrons within the atoms which compose the substance. In this case, the electrons will absorb the energy of the light wave and increase their energy state, often moving outward from the nucleus of the atom into an outer shell or orbital.

The atoms that bind together to make the molecules of any particular substance contain a number of electrons (given by the atomic number Z in the periodic chart). Recall that all light waves are electromagnetic in origin. Thus they are affected strongly when coming into contact with negatively charged electrons in matter. When photons (individual packets of light energy) come in contact with the valence electrons of atom, one of several things can and will occur:

- An electron absorbs all of the energy of the photon and re-emits it with different colour. This gives rise to luminescence, fluorescence and phosphorescence.
- An electron absorbs the energy of the photon and sends it back out the way it came in. This results in reflection or scattering.
- An electron cannot absorb the energy of the photon and the photon continues on its path. This results in transmission (provided no other absorption mechanisms are active).
- An electron selectively absorbs a portion of the photon, and the remaining frequencies are transmitted in the form of spectral colour.

Most of the time, it is a combination of the above that happens to the light that hits an object. The electrons in different materials vary in the range of energy that they can absorb. Most glasses, for example, block ultraviolet (UV) light. What happens is the electrons in the glass absorb the energy of the photons in the UV range while ignoring the weaker energy of photons in the visible light spectrum.

Thus, when a material is illuminated, individual photons of light can make the valence electrons of an atom transition to a higher electronic energy level. The photon is destroyed in the process and the absorbed radiant energy is transformed to electric potential energy. Several things can happen then to the absorbed energy. as it may be re-emitted by the electron as radiant energy (in this case the overall effect is in fact a scattering of light), dissipated to the rest of the material (i.e. transformed into heat), or the electron can be freed from the atom (as in the photoelectric and Compton effects).

### *Infrared: Bond Stretching*

The primary physical mechanism for storing mechanical energy of motion in condensed matter is through heat, or thermal energy. Thermal energy manifests itself as energy of motion. Thus, heat is motion at the atomic and molecular levels. The primary mode of motion in crystalline substances is vibration. Any given atom will vibrate around some mean or average position within a crystalline structure, surrounded by its nearest neighbours. This vibration in 2-dimensions is equivalent to the oscillation of a clock's pendulum. It swings back and forth symmetrically about some mean or average (vertical) position. Atomic and molecular vibrational frequencies may average on the order of $10^{12}$ cycles per second (hertz).

When a light wave of a given frequency strikes a material with particles having the same or (resonant) vibrational frequencies, then those particles will absorb the energy of the light wave and transform it into thermal energy of vibrational motion. Since different atoms and molecules have different natural frequencies of vibration, they will selectively absorb different frequencies (or portions of the spectrum) of infrared light. Reflection and transmission of light waves occur because the frequencies of the light waves do not match the natural resonant frequencies of vibration of the objects. When infrared light of these frequencies strikes an object, the energy is reflected or transmitted.

If the object is transparent, then the light waves are passed on to neighbouring atoms through the bulk of the material and re-emitted on the opposite side of the object. Such frequencies of light waves are said to be transmitted.

### *Transparency in Insulators*

An object may be not transparent either because it reflects the incoming light or because it absorbs the incoming light. Almost all solids reflect a part and absorb a part of the incoming light.

When light falls onto a block of metal, it encounters atoms that are tightly packed in a regular lattice and a "sea of electrons" moving randomly between the atoms. In metals, most of these are non-bonding electrons (or free electrons) as opposed to the bonding electrons typically found in covalently bonded or ionically bonded non-metallic (insulating) solids. In a metallic bond, any potential bonding electrons can easily be lost by the atoms in a crystalline structure. The effect of this delocalization is simply to exaggerate the effect of the "sea of electrons". As a result of these electrons, most of the incoming light in metals is reflected back, which is why we see a shiny metal surface.

Most insulators (or dielectric materials) are held together by ionic bonds. Thus, these materials do not have free conduction electrons, and the bonding electrons reflect only a small fraction of the incident wave. The remaining frequencies (or wavelengths) are free to propagate (or be transmitted). This class of materials includes all ceramics and glasses.

If a dielectric material does not include light-absorbent additive molecules (pigments, dyes, colourants), it is usually transparent to the spectrum of visible light. Colour centres (or dye molecules, or "dopants") in a dielectric absorb a portion of the incoming light wave. The remaining frequencies (or wavelengths) are free to be reflected or transmitted. This is how coloured glass is produced.

Most liquids and aqueous solutions are highly transparent. For example, water, cooking oil, rubbing alcohol, air, natural gas, are all clear. Absence of structural defects (voids, cracks, etc.) and molecular structure of most liquids are chiefly responsible for their excellent optical transmission. The ability of liquids to "heal" internal defects via viscous flow is one of the reasons why some fibrous materials (e.g., paper or fabric) increase their apparent transparency when wetted. The liquid fills up numerous voids making the material more structurally homogeneous.

Light scattering in an ideal defect-free crystalline (non-metallic) solid which provides *no scattering centres* for incoming lightwaves will be due primarily to any effects of anharmonicity within the ordered lattice. Lightwave transmission will be highly directional due to the typical anisotropy of crystalline substances, which includes their symmetry group and Bravais lattice. For example, the seven different crystalline forms of quartz silica (silicon dioxide, $SiO_2$) are all clear, transparent materials.

### Optical Waveguides

Optically transparent materials focus on the response of a material to incoming light waves of a range of wavelengths. Guided light wave transmission via frequency selective waveguides involves the emerging field of fibre optics and the ability of certain glassy compositions to act as a transmission medium for a range of frequencies simultaneously (multi-mode optical fibre) with little or no interference between competing wavelengths or frequencies. This resonant mode of energy and data transmission via electromagnetic (light) wave propagation is relatively lossless.

An optical fibre is a cylindrical dielectric waveguide that transmits light along its axis by the process of total internal reflection. The fibre consists of a core surrounded by a cladding layer. To confine the optical signal in the core, the refractive index of the core must be greater than that of the cladding. The refractive index is the parametre reflecting the speed of light in a material. (Refractive index is the ratio of the speed of light in vacuum to the speed of light in a given medium. The refractive index of vacuum is therefore 1.) The larger the refractive index, the more slowly light travels in that medium. Typical values for core and cladding of an optical fibre are 1.48 and 1.46, respectively.

When light travelling in a dense medium hits a boundary at a steep angle, the light will be completely reflected. This effect, called total internal reflection, is used in optical fibres to confine light in the core. Light travels along the fibre bouncing back and forth off of the boundary. Because the lıght must strike the boundary with an angle greater than the critical angle, only light that enters the fibre within a certain range of angles will be propagated. This range of angles is called the acceptance cone of the fibre. The size of this acceptance cone is a function of the refractive index difference between the fibre's core and cladding. Optical waveguides are used as components in integrated optical circuits (e.g. combined with lasers or light-emitting diodes, LEDs) or as the transmission medium in local and long haul optical communication systems.

## Mechanisms of Attenuation

Attenuation in fibre optics, also known as transmission loss, is the reduction in intensity of the light beam (or signal) with respect to distance travelled through a transmission medium. Attenuation coefficients in fibre optics usually use units of dB/km through the medium due to the very high quality of transparency of modern optical transmission media. The medium is usually a fibre of silica glass that confines the incident light beam to the inside. Attenuation is an important factor limiting the transmission of a signal across large distances. In optical fibres the main attenuation source is scattering from molecular level irregularities (Rayleigh scattering) due to structural disorder and compositional fluctuations of the glass structure. This same phenomenon is seen as one of the limiting factors in the transparency of infrared missile domes. Further attenuation is caused by light absorbed by residual materials, such as metals or water ions, within the fibre core and inner cladding. Light leakage due to bending, splices, connectors, or other outside forces are other factors resulting in attenuation.

### *As Camouflage*

Many marine animals that float near the surface are highly transparent, giving them almost perfect camouflage. However, transparency is difficult for bodies made of materials that have different refractive indices from seawater. Some marine animals such as jellyfish have gelatinous bodies, composed mainly of water; their thick mesogloea is acellular and highly transparent. This conveniently makes them buoyant, but it also makes them large for their muscle mass, so they cannot swim fast, making this form of camouflage a costly trade-off with mobility. Gelatinous planktonic animals are between 50 and 90 per cent transparent. A transparency of 50 per cent is enough to make an animal invisible to a predator such as cod at a depth of 650 metres (2,130 ft); better transparency is required for invisibility in shallower water, where the light is brighter and predators can see better. For example, a cod can see prey that are 98 per cent transparent in optimal lighting in shallow water. Therefore, sufficient transparency for camouflage is more easily achieved in deeper waters. For the same reason, transparency in air is even harder to achieve, but a partial example is found in the Glass frogs of the South American rain forest, which have translucent skin and pale greenish limbs.

## Streak (Mineralogy)

The streak (also called "powder colour") of a mineral is the colour of the powder produced when it is dragged across an unweathered surface. Unlike the apparent colour of a mineral, which for most minerals can vary considerably, the trail of finely ground powder generally has a more consistent characteristic colour, and is thus an important diagnostic tool in mineral identification. If no streak seems to be made, the mineral's streak is said to be white or colourless. Streak is particularly important as a diagnostic for opaque and coloured materials. It is less useful for silicate minerals, most of which have a white streak and are too hard to powder easily.

The apparent colour of a mineral can vary widely because of trace impurities or a disturbed macroscopic crystal structure. Small amounts of an impurity that strongly absorbs a particular wavelength can radically change the wavelengths of light that are reflected by the specimen, and thus change the apparent colour. However, when the specimen is dragged to produce a streak, it is broken into randomly oriented microscopic crystals, and small impurities do not greatly affect the absorption of light.

The surface across which the mineral is dragged is called a "streak plate," and is generally made of unglazed porcelain tile. In the absence of a streak plate, the unglazed underside of a porcelain bowl or vase or the back of a glazed tile will work. Sometimes a streak is more easily or accurately described by comparing it with the "streak" made by another streak plate.

Because the trail left behind results from the mineral being crushed into powder, a streak can only be made of minerals softer than the streak plate, around 7 on the Mohs scale of mineral hardness. In case of harder minerals, the colour of the powder can be determined by filing or crushing with a hammer a small sample, which is then usually rubbed on a streak plate. Most minerals that are harder have an unhelpful white streak.

Some minerals leave a streak similar to their natural colour, such as cinnabar and lazurite. Other minerals leave surprising colours, such as fluorite, which always has a white streak, although it can appear in purple, blue, yellow, or green crystals. Hematite, which is black in appearance, leaves a red streak which accounts for its name, which comes from the Greek word "haima," meaning "blood." Galena, which can be similar in appearance to hematite, is easily distinguished by its gray streak.

## Mohs Scale of Mineral Hardness

The Mohs scale of mineral hardness characterizes the scratch resistance of various minerals through the ability of a harder material to scratch a softer material. It was created in 1812 by the German geologist and mineralogist Friedrich Mohs and is one of several definitions of hardness in materials science. The method of comparing hardness by seeing which minerals can scratch others, however, is of great antiquity, having been mentioned by Theophrastus in his treatise *On Stones*, c. 300 BC, followed by Pliny the Elder in his *Naturalis Historia*, c. 77 AD.

### *Minerals*

The Mohs scale of mineral hardness is based on the ability of one natural sample of matter to scratch another mineral. The samples of matter used by Mohs are all different minerals. Minerals are pure substances found in nature. Rocks are made up of one or more minerals. As the hardest known naturally occurring substance when the scale was designed, diamonds are at the top of the scale. The hardness of a material is measured against the scale by finding the hardest material that the given material can scratch, and/or the softest material that can scratch the given material. For example, if some material is scratched by apatite but not by fluorite, its hardness on the Mohs scale would fall between 4 and 5.

The Mohs scale is a purely ordinal scale. For example, corundum (9) is twice as hard as topaz (8), but diamond (10) is four times as hard as corundum. The table below shows comparison with absolute hardness measured by a sclerometre, with pictorial examples.

| *Mohs hardness* | *Mineral* | *Chemical formula* | *Absolute hardness* |
|---|---|---|---|
| 1 | Talc | $Mg_3Si_4O_{10}(OH)_2$ | 1 |
| 2 | Gypsum | $CaSO_4 \cdot 2H_2O$ | 3 |
| 3 | Calcite | $CaCO_3$ | 9 |
| 4 | Fluorite | $CaF_2$ | 21 |
| 5 | Apatite | $Ca_5(PO_4)_3(OH^-,Cl^-,F^-)$ | 48 |
| 6 | Orthoclase Feldspar | $KAlSi_3O_8$ | 72 |
| 7 | Quartz | $SiO_2$ | 100 |
| 8 | Topaz | $Al_2SiO_4(OH^-,F^-)_2$ | 200 |
| 9 | Corundum | $Al_2O_3$ | 400 |
| 10 | Diamond | C | 1600 |

On the Mohs scale, graphite (a principal constituent of pencil "lead") has a hardness of 1.5; a fingernail, 2.2–2.5; a copper penny, 3.2–3.5; a pocketknife 5.1; a knife blade, 5.5; window glass plate, 5.5; and a steel file, 6.5. A streak plate (unglazed porcelain) has a hardness of 7.0. Using these ordinary materials of known hardness can be a simple way to approximate the position of a mineral on the scale.

### *Intermediate Hardness*

The table below incorporates additional substances that may fall between levels:

| Hardness | Substance or mineral |
|---|---|
| 0.2–0.3 | caesium, rubidium |
| 0.5–0.6 | lithium, sodium, potassium |
| 1 | talc |
| 1.5 | gallium, strontium, indium, tin, barium, thallium, lead, graphite |
| 2 | hexagonal boron nitride, calcium, selenium, cadmium, sulfur, tellurium, bismuth |
| 2.5–3 | magnesium, gold, silver, aluminium, zinc, lanthanum, cerium, Jet (lignite) |
| 3 | calcite, copper, arsenic, antimony, thorium, dentin |
| 4 | fluorite, iron, nickel |
| 4–4.5 | platinum, steel |
| 5 | apatite (tooth enamel), cobalt, zirconium, palladium, obsidian (volcanic glass) |
| 5.5 | beryllium, molybdenum, hafnium |
| 6 | orthoclase, titanium, manganese, germanium, niobium, rhodium, uranium |
| 6–7 | glass, fused quartz, iron pyrite, silicon, ruthenium, iridium, tantalum, opal, peridot |
| 7 | osmium, quartz, rhenium, vanadium |
| 7.5–8 | emerald, hardened steel, tungsten, spinel |
| 8 | topaz, cubic zirconia |
| 8.5 | chrysoberyl, chromium, silicon nitride, tantalum carbide |
| 9–9.5 | corundum, silicon carbide (carborundum), tungsten carbide, titanium carbide |
| 9.5–10 | boron, boron nitride, rhenium diboride, stishovite, titanium diboride |
| 10 | diamond, carbonado |
| >10 | nanocrystalline diamond (hyperdiamond, ultrahard fullerite) |

## *Hardness (Vickers)*

Comparison between Hardness (Mohs) and Hardness (Vickers):

| Mineralname | Hardness (Mohs) | Hardness (Vickers)kg/mm² |
|---|---|---|
| Graphite | 1–2 | $VHN_{10}$=7–11 |
| Tin | 1½ | $VHN_{10}$=7–9 |
| Bismuth | 2–2½ | $VHN_{100}$=16–18 |
| Gold | 2½ | $VHN_{10}$=30–34 |

*Contd...*

| ***Mineralname*** | ***Hardness (Mohs)*** | ***Hardness (Vickers)kg/mm²*** |
|---|---|---|
| Silver | 2½ | $VHN_{100}$=61–65 |
| Chalcocite | 2½–3 | $VHN_{100}$=84–87 |
| Copper | 2½–3 | $VHN_{100}$=77–99 |
| Galena | 2½ | $VHN_{100}$=79–104 |
| Sphalerite | 3½–4 | $VHN_{100}$=208–224 |
| Heazlewoodite | 4 | $VHN_{100}$=230–254 |
| Carrollite | 4½–5½ | $VHN_{100}$=507–586 |
| Goethite | 5–5½ | $VHN_{100}$=667 |
| Hematite | 5–6 | $VHN_{100}$=1,000–1,100 |
| Chromite | 5½ | $VHN_{100}$=1,278–1,456 |
| Anatase | 5½–6 | $VHN_{100}$=616–698 |
| Rutile | 6–6½ | $VHN_{100}$=894–974 |
| Pyrite | 6–6½ | $VHN_{100}$=1,505–1,520 |
| Bowieite | 7 | $VHN_{100}$=858–1,288 |
| Euclase | 7½ | $VHN_{100}$=1,310 |
| Chromium | 8½ | $VHN_{100}$=1,875–2,000 |

# 2

# Specific Gravity

Specific gravity is the ratio of the density of a substance to the density (mass of the same unit volume) of a reference substance. *Apparent* specific gravity is the ratio of the weight of a volume of the substance to the weight of an equal volume of the reference substance. The reference substance is nearly always water for liquids or air for gases. Temperature and pressure must be specified for both the sample and the reference. Pressure is nearly always 1 atm equal to 101.325 kPa. Temperatures for both sample and reference vary from industry to industry. In British brewing practice the specific gravity as specified above is multiplied by 1000. Specific gravity is commonly used in industry as a simple means of obtaining information about the concentration of solutions of various materials such as brines, hydrocarbons, sugar solutions (syrups, juices, honeys, brewers wort, must etc.) and acids.

## Details

Specific gravity, as it is a ratio of densities, is a dimensionless quantity. Specific gravity varies with temperature and pressure; reference and sample must be compared at the same temperature and pressure, or corrected to a standard reference temperature and pressure. Substances with a specific gravity of 1 are neutrally buoyant in water, those with SG greater than one are denser than water, and so (ignoring surface tension effects) will sink in it, and those with an SG of less than one are less dense than water, and so will float. In scientific work the relationship of mass to volume is usually expressed directly in terms of the density (mass per unit volume) of the substance under study. It is in industry where specific gravity finds wide application, often for historical reasons. True specific gravity can be expressed mathematically as:

$$SG_{\text{true}} = \frac{\rho_{\text{sample}}}{\rho_{H_2O}}$$

where $\rho_{\text{sample}}$ is the density of the sample and $\rho_{H_2O}$ is the density of water.

The apparent specific gravity is simply the ratio of the weights of equal volumes of sample and water in air:

$$SG_{\text{apparent}} = \frac{W_{A_{\text{sample}}}}{W_{A_{H_2O}}}$$

where $W_{A_{\text{sample}}}$ represents the weight of sample and $W_{A_{H_2O}}$ the weight of water, both measured in air.

It can be shown that true specific gravity can be computed from different properties:

$$SG_{\text{true}} = \frac{\rho_{\text{sample}}}{\rho_{H_2O}} = \frac{(m_{\text{sample}} / V)}{(m_{H_2O} / V)} = \frac{m_{\text{sample}}}{m_{H_2O}}\frac{g}{g} = \frac{W_{V_{\text{sample}}}}{W_{V_{H_2O}}}$$

where g is the local acceleration due to gravity, $V$ is the volume of the sample and of water (the same for both), $\rho_{\text{sample}}$ is the density of the sample, $\rho_{H_2O}$ is the density of water and $W_V$ represents a weight obtained in vacuum.

The density of water varies with temperature and pressure as does the density of the sample so that it is necessary to specify the temperatures and pressures at which the densities or weights were determined. It is nearly always the case that measurements are made at nominally 1 atmosphere (1013.25 mbar ± the variations caused by changing weather patterns) but as specific gravity usually refers to highly incompressible aqueous solutions or other incompressible substances (such as petroleum products) variations in density caused by pressure are usually neglected at least where apparent specific gravity is being measured. For true (*in vacuo*) specific gravity calculations air pressure must be considered. Temperatures are specified by the notation $(T_s / T_r)$ with $T_s$ representing $T_r$ the temperature at which the sample's density was determined and the temperature at which the reference (water) density is specified. For example SG (20°C/4°C) would be understood to mean that the density of the sample was determined at 20 °C and of the water at 4°C. Taking into account different sample and reference temperatures we note that $SG_{H_2O} = 1.000000$ while (20°C/20°C) it is also the case that $SG_{H_2O} = 0.998203 / 0.999840 = 0.998363$ (20°C/4°C). Here temperature is being specified using the current ITS-90 scale and the densities used here and in the rest of this article are based on that scale. On the previous IPTS-68 scale the densities at 20 °C and 4 °C are, respectively, 0.9982071 and 0.9999720 resulting in an SG (20°C/4°C) value for water of 0.9982343.

As the principal use of specific gravity measurements in industry is determination of the concentrations of substances in aqueous solutions and these are found in tables of SG vs concentration it is extremely important that the analyst enter the table with the correct form of specific gravity. For example, in the brewing industry, the Plato table, which lists sucrose concentration by weight against true SG, were originally (20°C/4°C) i.e. based on measurements of the density of sucrose solutions made at laboratory temperature (20 °C) but referenced to the density of water at 4 °C which is very close to the temperature at which water has its maximum density of $\rho_{H_2O}$ equal to 0.999972 $g \cdot cm^{-3}$ in SI units (or 62.43 $lb_m \cdot ft^{-3}$ in United States customary units). The ASBC table in use today in North America, while it is derived from the original Plato table is for apparent specific gravity measurements at (20°C/20°C) on the IPTS-68 scale where the density of water is 0.9982071 $g \cdot cm^{-3}$. In the sugar, soft drink, honey, fruit juice and related industries sucrose concentration by weight is taken from a table prepared by A. Brix which uses SG (17.5°C/17.5°C). As a final example, the British SG units are based on reference and sample temperatures of 60F and are thus (15.56°C/15.56°C).

Given the specific gravity of a substance, its actual density can be calculated by rearranging the above formula:

$$\rho_{\text{substance}} = SG \times \rho_{H_2O}.$$

Occasionally a reference substance other than water is specified (for example, air), in which case specific gravity means density relative to that reference.

## *Measurement: Apparent and True Specific Gravity*

***Pycnometre:*** Specific gravity can be measured in a number of ways. The following illustration involving the use of the pycnometre is instructive. A pycnometre is simply a bottle which can be precisely filled to a specific, but not necessarily accurately known volume, $V$. Placed upon a balance of some sort it will exert a force.

$$F_b = g(m_b - \rho_a \frac{m_b}{\rho_b})$$

where $m_b$ is the mass of the bottle and g the gravitational acceleration at the location at which the measurements are being made. $\rho_a$ is the density of the air at the ambient pressure and $\rho_b$ is the density of the material of which the bottle is made (usually glass) so that the second term is the mass of air displaced by the glass of the bottle whose weight, by Archimedes Principle must be subtracted. The bottle is, of course, filled with air but as that air displaces an equal amount of air the weight of that air is cancelled by the weight of the air displaced. Now we fill the bottle with the reference fluid e.g. pure water. The force exerted on the pan of the balance becomes:

$$F_w = g(m_b - \rho_a \frac{m_b}{\rho_b} + V\rho_w - V\rho_a).$$

If we subtract the force measured on the empty bottle from this (or tare the balance before making the water measurement) we obtain.

$$F_{w,n} = gV(\rho_w - \rho_a)$$

where the subscript n indicated that this force is net of the force of the empty bottle. The bottle is now emptied, thoroughly dried and refilled with the sample. The force, net of the empty bottle, is now:

$$F_{s,n} = gV(\rho_s - \rho_a)$$

where $\rho_s$ is the density of the sample. The ratio of the sample and water forces is:

$$SG_A = \frac{gV(\rho_s - \rho_a)}{gV(\rho_w - \rho_a)} = \frac{(\rho_s - \rho_a)}{(\rho_w - \rho_a)}.$$

This is called the Apparent Specific Gravity, denoted by subscript A, because it is what we would obtain if we took the ratio of net weighings in air from an analytical balance or used a hydrometre (the stem displaces air). Note that the result does not depend on the calibration of the balance. The only requirement on it is that it read linearly with force. Nor does $SG_A$ depend on the actual volume of the pycnometre.

Further manipulation and finally substitution of $SG_V$ ,the true specific gravity, (the subscript V is used because this is often referred to as the specific gravity *in vacuo*) for $\frac{\rho_s}{\rho_w}$ gives the relationship between apparent and true specific gravity.

$$SG_A = \frac{\frac{\rho_s}{\rho_w} - \frac{\rho_a}{\rho_w}}{1 - \frac{\rho_a}{\rho_w}} = \frac{SG_V - \frac{\rho_a}{\rho_w}}{1 - \frac{\rho_a}{\rho_w}}$$

In the usual case we will have measured weights and want the true specific gravity. This is found from

$$SG_V = SG_A - \frac{\rho_a}{\rho_w}(SG_A - 1).$$

Since the density of dry air at 1013.25 mb at 20 °C is 0.001205 g·cm$^{-3}$ and that of water is 0.998203 g·cm$^{-3}$ the difference between true and apparent specific gravities for a substance with specific gravity (20°C/20°C) of about 1.100 would be 0.000120. Where the specific gravity of the sample is close to that of water (for example dilute ethanol solutions) the correction is even smaller.

## Digital Density Metres

***Hydrostatic Pressure-based Instruments*:** This technology relies upon Pascal's Principle which states that the pressure difference between two points within a vertical column of fluid is dependent upon the vertical distance between the two points, the density of the fluid and the gravitational force. This technology is often used for tank gauging applications as a convenient means of liquid level and density measure.

***Vibrating Element Transducers*:** This type of instrument requires a vibrating element to be placed in contact with the fluid of interest. The resonant frequency of the element is measured and is related to the density of the fluid by a characterization that is dependent upon the design of the element. In modern laboratories precise measurements of specific gravity are made using oscillating U-tube metres. These are capable of measurement to 5 to 6 places beyond the decimal point and are used in the brewing, distilling, pharmaceutical, petroleum and other industries. The instruments measure the actual mass of fluid contained in a fixed volume at temperatures between 0 and 80 °C but as they are microprocessor based can calculate apparent or true specific gravity and contain tables relating these to the strengths of common acids, sugar solutions, etc. The vibrating fork immersion probe is another good example of this technology. This technology also includes many coriolis-type mass flow metres which are widely used in chemical and petroleum industry for high accuracy mass flow measurement and can be configured to also output density information based on the resonant frequency of the vibrating flow tubes.

***Ultrasonic Transducer*:** Ultrasonic waves are passed from a source, through the fluid of interest, and into a detector which measures the acoustic spectroscopy of the waves. Fluid properties such as density and viscosity can be inferred from the spectrum.

***Radiation-based Gauge:*** Radiation is passed from a source, through the fluid of interest, and into a scintillation detector, or counter. As the fluid density increases, the detected radiation "counts" will decrease. The source is typically the radioactive isotope cesium-137, with a half-life of about 30 years. A key advantage for this technology is that the instrument is not required to be in contact with the fluid – typically the source and detector are mounted on the outside of tanks or piping.

***Buoyant Force Transducer*:** the buoyancy force produced by a float in a homogeneous liquid is equal to the weight of the liquid that is displaced by the float. Since buoyancy force is linear with respect to the density of the liquid within which the float is submerged, the measure of the buoyancy force yields a measure of the density of the liquid. One commercially available unit claims the instrument is capable of measuring specific gravity with an accuracy of +/- 0.005 SG units. The submersible probe head contains a mathematically characterized spring-float system. When the head is immersed vertically in the liquid, the float moves vertically and the position of the float controls the position of a permanent magnet whose displacement is sensed by a concentric array of Hall-effect linear displacement sensors. The output signals of the sensors are mixed in a dedicated electronics module that provides an output voltage whose magnitude is a direct linear measure of the quantity to be measured.

***In-Line Continuous Measurement*:** Slurry is weighed as it travels through the metreed section of pipe using a patented, high resolution load cell. This section of pipe is of optimal length such that a truly representative mass of the slurry may be determined. This representative mass is then interrogated by the load cell 110 times per second to ensure accurate and repeatable measurement of the slurry.

### *Examples*

- Helium gas has a density of 0.164g/litre It is 0.139 times as dense as air.
- Air has a density of 1.18g/l

| *Material* | *Specific Gravity* |
|---|---|
| Balsa wood | 0.2 |
| Oak wood | 0.75 |
| Ethanol | 0.78 |
| Water | 1 |
| Table salt | 2.17 |
| Aluminium | 2.7 |
| Iron | 7.87 |
| Copper | 8.96 |
| Lead | 11.35 |
| Mercury | 13.56 |
| Depleted uranium | 19.1 |
| Gold | 19.3 |
| Osmium | 22.59 |

*(Samples may vary, and these figures are approximate.)*

- Urine normally has a specific gravity between 1.003 and 1.035.
- Blood normally has a specific gravity of ~1.060.

## Miller Index

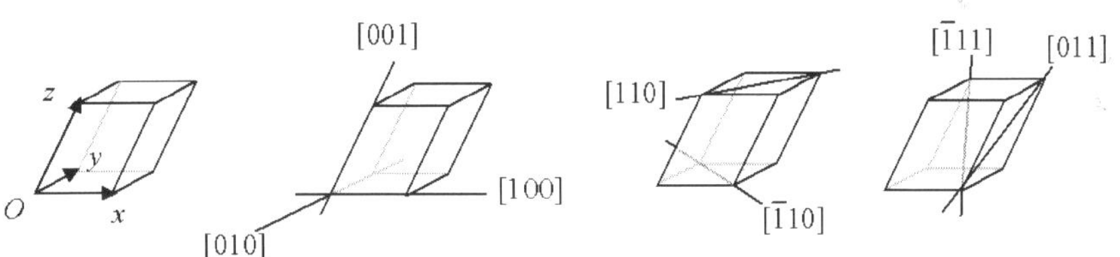


***Figure:*** *Examples of directions*

Miller indices form a notation system in crystallography for planes in crystal (Bravais) lattices.

In particular, a family of lattice planes is determined by three integers *h*, *k*, and $\ell$, the *Miller indices.* They are written (hk$\ell$), and each index denotes a plane orthogonal to a direction (h, k, $\ell$) in the basis of the reciprocal lattice vectors. By convention, negative integers are written with a bar, as in 3 for • 3. The integers are usually written in lowest terms, i.e. their greatest common divisor should be 1. Miller index 100 represents a plane orthogonal to direction *h*; index 010 represents a plane orthogonal to direction *k*, and index 001 represents a plane orthogonal to $\ell$.

There are also several related notations:

- the notation {hk$\ell$} denotes the set of all planes that are equivalent to (hk$\ell$) by the symmetry of the lattice.

In the context of crystal directions (not planes), the corresponding notations are:

- [hk$\ell$], with square instead of round brackets, denotes a direction in the basis of the *direct* lattice vectors instead of the reciprocal lattice; and
- similarly, the notation $\langle hk\ell \rangle$ denotes the set of all directions that are equivalent to [hk$\ell$] by symmetry.

Miller indices were introduced in 1839 by the British mineralogist William Hallowes Miller. The method was also historically known as the Millerian system, and the indices as Millerian, although this is now rare.

The Miller indices are defined with respect to any choice of unit cell and not only with respect to primitive basis vectors, as is sometimes stated.

### Definition

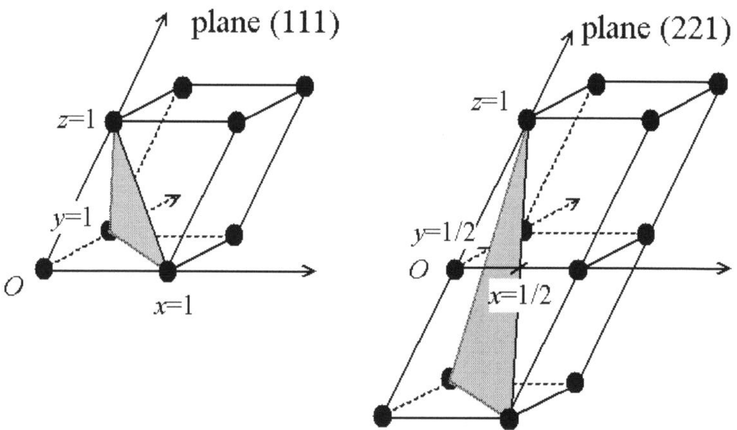


***Figure:*** *Examples of determining indices for a plane using intercepts with axes; left (111), right (221)*

There are two equivalent ways to define the meaning of the Miller indices: via a point in the reciprocal lattice, or as the inverse intercepts along the lattice vectors. Both definitions are given below. In either case, one needs to choose the three lattice vectors $a_1$, $a_2$, and $a_3$ that define the unit cell (note that the conventional unit cell may be larger than the primitive cell of the Bravais lattice, as the examples below illustrate). Given these, the three primitive reciprocal lattice vectors are also determined (denoted $b_1$, $b_2$, and $b_3$).

Then, given the three Miller indices h, k, $\ell$, (hk$\ell$) denotes planes orthogonal to the reciprocal lattice vector:

$$g_{hk\ell} = hb_1 + kb_2 + \ell b_3.$$

That is, (hk$\ell$) simply indicates a normal to the planes in the basis of the primitive reciprocal lattice vectors. Because the coordinates are integers, this normal is itself always a reciprocal lattice vector. The requirement of lowest terms means that it is the *shortest* reciprocal lattice vector in the given direction.

Equivalently, (hk$\ell$) denotes a plane that intercepts the three points $a_1$/h, $a_2$/k, and $a_3$/$\ell$, or some multiple thereof. That is, the Miller indices are proportional to the *inverses* of the intercepts of the plane, in the basis of the lattice vectors. If one of the indices is zero, it means that the planes do not intersect that axis (the intercept is "at infinity").

Considering only (hk$\ell$) planes intersecting one or more lattice points (the *lattice planes*), the perpendicular distance *d* between adjacent lattice planes is related to the (shortest) reciprocal lattice vector orthogonal to the planes by the formula: $d = 2\pi / |g_{hk\ell}|$.

The related notation [hk$\ell$] denotes the *direction*:

$$ha_1 + ka_2 + \ell a_3.$$

That is, it uses the direct lattice basis instead of the reciprocal lattice. Note that [hk$\ell$] is *not* generally normal to the (hk$\ell$) planes, except in a cubic lattice as described below.

### Case of Cubic Structures

For the special case of simple cubic crystals, the lattice vectors are orthogonal and of equal length (usually denoted *a*); similar to the reciprocal lattice. Thus, in this common case, the Miller indices (hk$\ell$) and [hk$\ell$] both simply denote normals/directions in Cartesian coordinates.

For cubic crystals with lattice constant *a*, the spacing *d* between adjacent (hk$\ell$) lattice planes is (from above):

$$d_{hk\ell} = \frac{a}{\sqrt{h^2 + k^2 + \ell^2}}.$$

Because of the symmetry of cubic crystals, it is possible to change the place and sign of the integers and have equivalent directions and planes:

- Coordinates in *angle brackets* such as ⟨100⟩ denote a *family* of directions which are equivalent due to symmetry operations, such as [100], [010], [001] or the negative of any of those directions.
- Coordinates in *curly brackets* or *braces* such as {100} denote a family of plane normals which are equivalent due to symmetry operations, much the way angle brackets denote a family of directions.

For face-centred cubic and body-centred cubic lattices, the primitive lattice vectors are not orthogonal. However, in these cases the Miller indices are conventionally defined relative to the lattice vectors of the cubic supercell and hence are again simply the Cartesian directions.

### Case of Hexagonal and Rhombohedral Structures

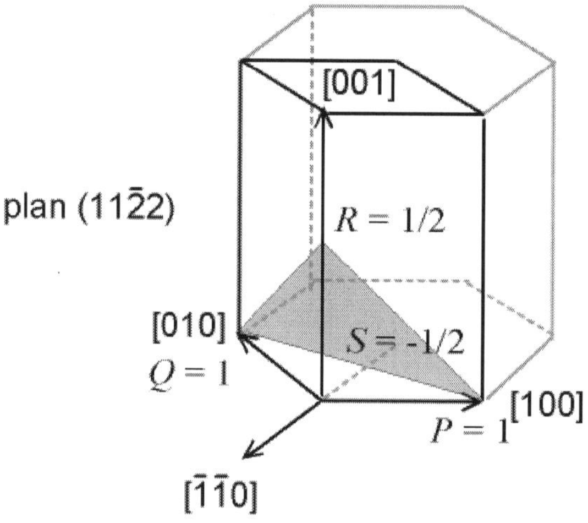


***Figure:*** *Miller-Bravais indices*

With hexagonal and rhombohedral lattice systems, it is possible to use the Bravais-Miller index which has 4 numbers ($h$ $k$ $i$ $\ell$)

$i = -(h + k)$.

Here $h$, $k$ and $\ell$ are identical to the Miller index, and $i$ is a redundant index.

This four-index scheme for labelling planes in a hexagonal lattice makes permutation symmetries apparent. For example, the similarity between (110) ≡ (1120) and (120) ≡ (1210) is more obvious when the redundant index is shown.

In the figure at right, the (001) plane has a 3-fold symmetry: it remains unchanged by a rotation of 1/3 (2π/3 rad, 120°). The [100],

[010] and the [110] directions are really similar. If $S$ is the intercept of the plane with the [110] axis, then

$i = 1/S$.

There are also *ad hoc* schemes (e.g. in the transmission electron microscopy literature) for indexing hexagonal *lattice vectors* (rather than reciprocal lattice vectors or planes) with four indices. However they don't operate by similarly adding a redundant index to the regular three-index set.

For example, the reciprocal lattice vector (hk$\ell$) as suggested above can be written as ha*+kb*+$\ell$c*if the reciprocal-lattice basis-vectors are a*, b*, and c*. For hexagonal crystals this may be expressed in terms of direct-lattice basis-vectors a, b and c as

$$(hk\ell) = h\vec{a}^* + k\vec{b}^* + \ell\vec{c}^* = \frac{2}{3a^2}(2h+k)\vec{a} + \frac{2}{3a^2}(h+2k)\vec{b} + \frac{1}{c^2}(\ell)\vec{c}.$$

Hence zone indices of the direction perpendicular to plane (hk$\ell$) are, in suitably normalized triplet form, simply [2h+k,h+2k, $\ell$(3/2)(a/c)$^2$]. When *four indices* are used for the zone normal to plane (hk$\ell$), however, the literature often uses [h,k,-h-k, $\ell$(3/2)(a/c)$^2$] instead. Thus as you can see, four-index zone indices in square or angle brackets sometimes mix a single direct-lattice index on the right with reciprocal-lattice indices (normally in round or curly brackets) on the left.

### *Integer vs. Irrational Miller Indices: Lattice Planes and Quasicrystals*

Ordinarily, Miller indices are always integers by definition, and this constraint is physically significant. To understand this, suppose that we allow a plane (abc) where the Miller "indices" $a$, $b$ and $c$ (defined as above) are not necessarily integers.

If $a$, $b$ and $c$ have rational ratios, then the same family of planes can be written in terms of integer indices (hk$\ell$) by scaling $a$, $b$ and $c$ appropriately: divide by the largest of the three numbers, and then multiply by the least common denominator. Thus, integer Miller indices implicitly include indices with all rational ratios. The reason why planes where the components (in the reciprocal-lattice basis) have rational ratios are of special interest is that these are the lattice planes: they are the only planes whose intersections with the crystal are 2d-periodic.

For a plane (abc) where $a$, $b$ and $c$ have irrational ratios, on the other hand, the intersection of the plane with the crystal is *not* periodic. It forms an aperiodic pattern known as a quasicrystal. This construction corresponds precisely to the standard "cut-and-project" method of defining a quasicrystal, using a plane with irrational-ratio Miller indices. (Although many quasicrystals, such as the Penrose tiling, are formed by "cuts" of periodic lattices in more than three dimensions, involving the intersection of more than one such hyperplane.)

### *Chemical Mineralogy*

Chemical mineralogy focuses on the chemical composition of minerals in order to identify, classify, and categorize them, as well as a means to find beneficial uses from them. There are a few minerals which are classified as whole elements, including sulfur, copper, silver, and gold, yet the vast majority of minerals are chemical compounds, some more complex than others. In terms of major chemical divisions of minerals, most are placed within the isomorphous groups, which are based on analogous chemical composition and similar crystal forms. A good example of isomorphism classification would be the calcite group, containing the minerals calcite, magnesite, siderite, rhodochrosite, and smithsonite.

### *Biomineralogy*

Biomineralogy is a cross-over field between mineralogy, paleontology and biology. It is the study of how plants and animals stabilize minerals under biological control, and the sequencing of mineral replacement of those minerals after deposition. It uses techniques from chemical mineralogy, especially isotopic studies, to determine such things as growth forms in living plants and animals as well as things like the original mineral content of fossils.

## Refractive Index

In optics the refractive index or index of refraction $n$ of a substance (optical medium) is a dimensionless number that describes how light, or any other radiation, propagates through that medium. It is defined as where $c$ is the speed of light in vacuum and $v$ is the speed of light in the substance. For example, the refractive index of water is 1.33, meaning that light travels 1.33 times slower in water than it does in vacuum.

The historically first occurrence of the refractive index was in Snell's law of refraction, $n_1 \sin\theta_1 = n_2 \sin\theta_2$, where $\theta_1$ and $\theta_2$ are the angles of incidence of a ray crossing the interface between two media with refractive indices $n_1$ and $n_2$.

Brewster's angle, the critical angle for total internal reflection, and the reflectivity of a surface also depend on the refractive index, as described by the Fresnel equations. The refractive index can be seen as the factor by which the velocity and the wavelength of the radiation are reduced with respect to their vacuum values: the speed of light in a medium is $v = c / n$ and similarly the wavelength in that medium is $\lambda = \lambda_0 / n$, where $\lambda_0$ is the wavelength of that light in vacuum. This implies that vacuum has a refractive index of 1. Historically other reference media (e.g., air at a standardized pressure and temperature) have been common.

Refractive index of materials varies with the wavelength. This is called dispersion; it causes the splitting of white light in prisms and rainbows, and chromatic aberration in lenses. In opaque media, the refractive index is a complex number: while the real part describes refraction, the imaginary part accounts for absorption.

The concept of refractive index is widely used within the full electromagnetic spectrum, from x-rays to radio waves. It can also be used with wave phenomena other than light (e.g., sound). In this case the speed of sound is used instead of that of light and a reference medium other than vacuum must be chosen.

### *History*

Thomas Young was presumably the person who first used, and invented, the name "index of refraction", in 1807. At the same time he changed this value of refractive power into a single number, instead of the traditional ratio of two numbers. The ratio had the disadvantage of different appearances. Newton, who called it the "proportion of the sines of incidence and refraction", wrote it as a ratio of two numbers, like "529 to 396" (or "nearly 4 to 3"; for water). Hauksbee, who called it the "ratio of refraction", wrote it as a ratio with a fixed numerator, like "10000 to 7451.9" (for urine). Hutton wrote it as a ratio with a fixed denominator, like 1.3358 to 1 (water).

Young did not use a symbol for the index of refraction, in 1807. In the next years, others started using different symbols: n, m, and $\mu$. The symbol n gradually prevailed.

### *Refractive Index Below 1*

A widespread misconception is that since, according to the theory of relativity, nothing can travel faster than the speed of light in vacuum, the refractive index cannot be lower than 1. This is erroneous since the refractive index measures the phase velocity of light, which does not carry information. The phase velocity is the speed at which the crests of the wave move and can be faster than the speed of light in vacuum, and thereby give a refractive index below 1. This can occur close to resonance frequencies, for absorbing media, in plasmas, and for x-rays. In the x-ray regime the refractive indices are lower than but very close to 1 (exceptions close to some resonance frequencies). As an example, water has a refractive index of $1 - 2.6\times10^{-7}$ for X-ray radiation at a photon energy of 30 keV (0.04 nm wavelength).

## Negative Index Metamaterials

Negative index metamaterial or negative index material (NIM) is an artificial structure where the refractive index has a negative value over some frequency range. This does not occur in any known natural materials, and thus is only achievable with engineered structures known as *metamaterials.* Metamaterial refers broadly to any synthetic material with unusual refractive properties.

NIMs are constructed of the same manufactured basic parts called unit cells. For example, the unit cells of the first NIMs were constructed from circuit board material, or in other words, wires and dielectrics. And in general, these artificially constructed cells are stacked or planar and configured in a particular repeated pattern to compose the individual NIM material. For instance, the unit cells of the first NIMs were stacked horizontally and vertically, resulting in a pattern that was repeated and intended.

Specifications for the response of each unit cell are predetermined prior to construction and are based on the intended or desired response of the entire, newly constructed, material. In other words, each cell is individually tuned to respond in a certain way, based on the desired output of the NIM. The aggregate response of the NIM is unlike the response of its constituent materials, at both the granular and cellular level, and this aggregate response is not likely to be produced in nature. In other words, the way the NIM responds is that of a new material, unlike the wires and dielectrics it is made from, and unlike the cells that are also made from these materials.

The aggregate response of the NIM is the sought after result and is in effect an average, albeit a complicated and non-linear average. Hence, the NIM has become an effective medium. Also, in effect, this metamaterial has become an "ordered macroscopic material, synthesized from the bottom up", and has emergent properties beyond its components.

Metamaterials which exhibit a negative value for the refractive index such as negative index materials are often referred to by any of several terminologies: left-handed media or left-handed material (LHM), backward wave media (BW media), media with negative refractive index, double negative (DNG) metamaterials, and other similar names.

### Properties and Characteristics

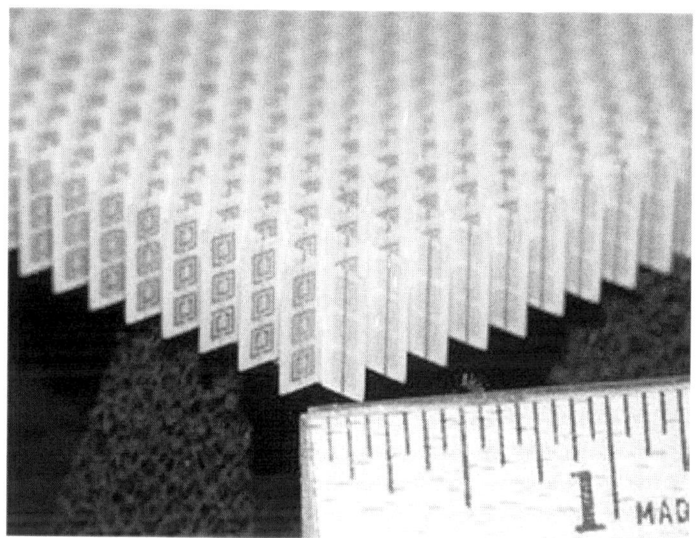


***Figure:*** *A split-ring resonator array arranged to produce a negative index of refraction, constructed of copper split-ring resonators and wires mounted on interlocking sheets of fibreglass circuit board.*

The total array consists of 3 by 20×20 unit cells with overall dimensions of 10×100×100 milimetres. The height of 10 milimetres measures a little more than six subdivision marks on the ruler, which is marked in inches.

Credit: NASA Glenn Research Centre.

Metamaterials were first proposed by a Russian theorist Victor Veselago in 1967. The proposed *left-handed* or *negative index* materials were theorized to exhibit optical properties opposite to those of glass, air, and other transparent media. Such materials were predicted to exhibit counterintuitive properties like bending or refracting light in unusual and unnatural ways. However, the first practical metamaterial was not constructed until 33 years later and it does produce Veselago's concepts.

Currently, negative index metamaterials are being developed to manipulate electromagnetic radiation in new ways. For example, optical and electromagnetic properties of natural materials are often altered through chemistry. With metamaterials optical and electromagnetic properties can be engineered by changing the geometry of its unit cells. The unit cells are materials that are ordered in geometric arrangements with dimensions that are fractions of the wavelength of the radiated electromagnetic wave. Each artificial unit responds to the radiation from the source. The collective result is the material's response to the electromagnetic wave that is broader than normal.

Subsequently, transmission is altered by adjusting the shape, size, and configurations of the unit cells. This results in control over material parametres known as permittivity and magnetic permeability. These two parametres (or quantities) determine the propagation of electromagnetic waves in matter. Therefore, controlling the values of permittivity and permeability means that the refractive index can be negative or zero as well as conventionally positive. It all depends on the intended application or desired result. So, optical properties can be expanded beyond the capabilities of lenses, mirrors, and other conventional materials. Additionally, one of the effects most studied is the negative index of refraction.

### *Reverse Propagation*

When a negative index of refraction occurs, propagation of the electromagnetic wave is reversed. Resolution below the diffraction limit becomes possible. This is known as Subwavelength imaging. Transmitting a beam of light via an electromagnetically flat surface is another capability. In contrast, conventional materials are usually curved, and cannot achieve resolution below the diffraction limit. Also, reversing the electromagnetic waves in a material, in conjunction with other ordinary materials (including air) could result in minimizing losses that would normally occur.

The reverse of the electromagnetic wave, characterized by an antiparallel phase velocity is also an indicator of negative index of refraction.

Furthermore, negative index materials are customized composites. In other words, materials are combined with a desired result in mind. Combinations of materials can be designed to achieve optical properties not seen in nature. The properties of the composite material stem from its lattice structure constructed from components smaller than the impinging electromagnetic wavelength separated by distances that are also smaller than the impinging electromagnetic wavelength. Likewise, by fabricating such metamaterials researchers are trying to overcome fundamental limits tied to the wavelength of light. The unusual and counter intuitive properties currently have practical and commercial use manipulating electromagnetic microwaves in wireless and communication systems. Lastly, research continues in the other domains of the electromagnetic spectrum, including visible light.

### *Resonance*

Negative index metamaterial response is usually linked to the resonant behaviour of the unit cells. However, transmission line components that use metamaterials may employ a different technique.

### *Materials*

The first actual metamaterials worked in the microwave regime, or centimetre wavelengths, of the electromagnetic spectrum (about 4.3 GHz). It was constructed of split-ring resonators and conducting straight wires (as unit cells). The unit cells were sized from 7 to 10 millimetres. The unit cells were arranged in a two-dimensional (periodic) repeating pattern which produces a crystal-like geometry. Both, the unit cells and the lattice spacing were smaller than the radiated electromagnetic wave. This produced the first left-handed material when both the permittivity and permeability of the material were negative. This system relies on the resonant behaviour of the unit cells. Below a group of researchers develop an idea for a left handed metamaterial that does not rely on such resonant behaviour.

Research in the microwave range continues with split-ring resonators and conducting wires. Research also continues in the shorter wavelengths with this configuration of materials and the unit cell sizes are scaled down. However, at around 200 terahertz issues arise which make using the split ring resonator problematic. "*Alternative materials become more suitable for the terahertz and optical regimes.*" At these wavelengths selection of materials and size limitations become important. For example, in 2007 a 100 nanometer mesh wire design made of silver and woven in a repeating pattern transmitted beams at the 780 nanometer wavelength, the far end of the visible spectrum. The researchers believe this produced a negative refraction of 0.6. Nevertheless, this operates at only a single wavelength like its predecessor metamaterials in the microwave regime. Hence, the challenges are to fabricate metamaterials so that they "refract light at ever-smaller wavelengths" and to develop broad band capabilities.

## *Artificial Transmission-line-media*

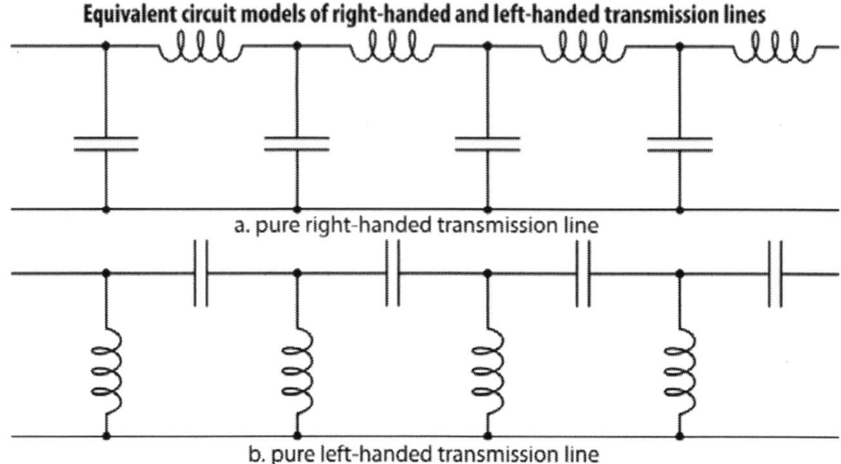


In the metamaterial literature, medium or media refers to transmission medium or optical medium. In 2002 a group of researchers came up with the idea that in contrast to materials that depended on resonant behaviour, non-resonant phenomena could surpass narrow bandwidth constraints of the wire/split-ring resonator configuration. This idea translated into a type of medium with broader bandwidth abilities, negative refraction, backward waves, and focusing beyond the diffraction limit.

They dispensed with split-ring-resonators and instead used a network of L–C loaded transmission lines. In the metamaterial literature this became known as artificial transmission-line media. At that time it had the added advantage of being more compact than a unit made of wires and split ring resonators. The network was both scalable (from the megahertz to the tens of gigahertz range) and tunable. It also includes a method for focusing the wavelengths of interest. By 2007 the negative refractive index transmission line was employed as a subwavelength focusing free-space flat lens. That this is a free-space lens is a significant advance. Part of prior research efforts targeted creating a lens that did not need to be embedded in a transmission line.

## *The Optical Domain*

Metamaterial components shrink as research explores shorter wavelengths (higher frequencies) of the electromagnetic spectrum in the infrared and visible spectrums. For example theory and experiment have investigated smaller horseshoe shaped split ring resonators designed with lithographic techniques, as well as paired metal nanorods or nanostrips, and nanoparticles as circuits designed with lumped element models

## Applications

The science of negative index materials is being matched with conventional devices that broadcast, transmit, shape, or receive electromagnetic signals that travel over cables, wires, or air. The materials, devices and systems that are involved with this work could have their properties altered or heightened. Hence, this is already happening with metamaterial antennas and related devices which are commercially available. Moreover, in the wireless domain these metamaterial apparatuses continue to be researched. Other applications are also being researched. These are electromagnetic absorbers such as radar-microwave absorbers, electrically small resonators, waveguides that can go beyond the diffraction limit, phase compensators, advancements in focusing devices (e.g. microwave lens), and improved electrically small antennas.

In the optical frequency regime developing the superlens may allow for imaging below the diffraction limit. Other potential applications for negative index metamaterials are optical nanolithography, nanotechnology circuitry, as well as a near field superlens (Pendry, 2000) that could be useful for biomedical imaging and subwavelength photolithography.

## Manipulating Permittivity and Permeability

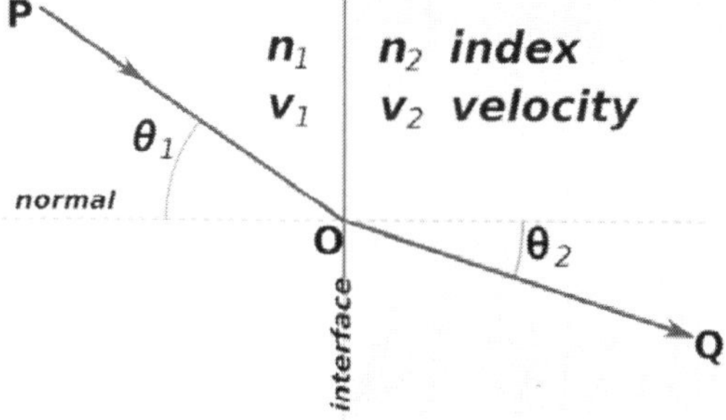


***Figure:*** *Refraction of light at the interface between two media of different refractive indices, with* $n_2 > n_1$. *Since the velocity is lower in the second medium* ($v_2 < v_1$), *the angle of refraction* $\theta_2$ *is less than the angle of incidence* $\theta_1$; *that is, the ray in the higher-index medium is closer to the normal.*

To describe any electromagnetic properties of a given material such as an optical lens, two significant parametres should be noted.

These are permittivity, ε, and permeability, μ, which could allow for accurate prediction of light waves travelling within materials, and electromagnetic phenomena that occur at the surface between two materials (interface).

For example, refractive index is an electromagnetic phenomenon which occurs at the surface (or interface) between two materials. Snell's law states that the relationship between the radiated angle of incidence, and the resulting refracted angle of transmission, rests on the refractive index, *n*, of the two media (materials). Mathematics provides a visualization with $n = \pm\sqrt{\epsilon\mu}$ . Hence, it can be seen that the behaviour of the refractive index is dependent on the association of these two parametres, as well as their quantitative values. Therefore, if designed or arbitrarily modified values can be inputs for ε, and, μ then the behaviour of propagating electromagnetic waves inside the material can be manipulated at will. This ability then allows for intentional determination of the refractive index.

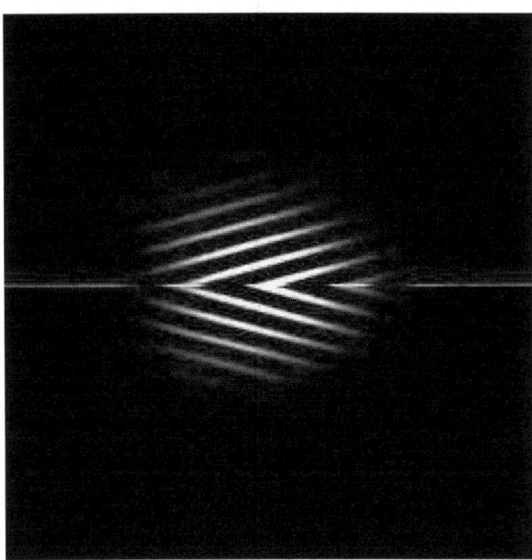

***Figure:*** *Video representing negative refraction of light at uniform planar interface.*

For example, in 1967, Victor Veselago analytically determined that light will refract in the reverse direction (negatively) at the interface between a material with negative refractive index and a material exhibiting conventional refractive index. This extraordinary material was realized on paper with simultaneous negative values for ε, and, μ, and could therefore be termed a double negative material. However, in Veselago's day a material which exhibits double negative parametres simultaneously seemed impossible because no natural materials exist which can produce this effect. Therefore his work was ignored for three decades.

In general the physical properties of natural materials cause limitations. Most dielectrics only have positive permittivities, $\varepsilon > 0$ . Metals will exhibit negative permittivity, $\varepsilon < 0$ at optical frequencies, and plasmas exhibit negative permittivity values in certain frequency bands. Pendry et al. demonstrated that the plasma frequency can be made to occur in the lower microwave frequencies for metals with a material made of metal rods that replaces the bulk metal. However, in each of these cases permeability remains always positive. At microwave frequencies it is possible for negative $\mu$ to occur in some ferromagnetic materials. But the inherent drawback is they are difficult to find above terahertz frequencies. In any case, a natural material that can achieve negative values for permittivity and permeability simultaneously has not been found or discovered. Hence, all of this has led to constructing artificial composite materials known as metamaterials in order to achieve the desired results.

### *Physical Properties Never Before Produced in Nature*

Theoretical articles were published in 1996 and 1999 which showed that synthetic materials could be constructed to purposely exhibit a negative permittivity and permeability.

These papers, along with Veselago's 1967 theoretical analysis of the properties of negative index materials, provided the background to fabricate a metamaterial with negative effective permittivity and permeability.

A metamaterial developed to exhibit negative index behaviour is typically formed from individual components. Each component responds independently and differently to a radiated electromagnetic wave as it travels through the material. Since these components are smaller than the radiated wavelength it is understood that a macroscopic view includes an effective value for both permittivity and permeability.

### *Composite Material*

In the year 2000, David R. Smith's team of UCSD researchers produced a new class of composite materials by depositing a structure onto a circuit-board substrate consisting of a series of thin copper split-rings and ordinary wire segments strung parallel to the rings. This material exhibited unusual physical properties that had never been observed in nature. These materials obey the laws of physics, but behave differently from normal materials. In essence these *negative index metamaterials* were noted for having the ability to reverse many of the physical properties that govern the behaviour of ordinary optical materials. One of those unusual properties is the ability to reverse, for the first time, Snell's law of refraction. Until the demonstration of negative refractive index for microwaves by the UCSD team, the material had been unavailable. Advances during the 1990s in fabrication and computation abilities allowed these first metamaterials to be constructed. Thus, the "new" metamaterial was tested for the effects described by Victor Veselago 30 years earlier. Studies of this experiment, which followed shortly thereafter, announced that other effects had occurred.

With antiferromagnets and certain types of insulating ferromagnets, effective negative magnetic permeability is achievable when polariton resonance exists. To achieve a negative index of refraction, however, permittivity with negative values must occur within the same frequency range. The artificially fabricated split-ring resonator is a design that accomplishes this, along with the promise of dampening high losses. With this first introduction of the metamaterial, it appears that the losses incurred were smaller than antiferromagnetic, or ferromagnetic materials.

When first demonstrated in 2000, the composite material (NIM) was limited to transmitting microwave radiation at frequencies of 4 to 7 gigahertz (4.28-7.49 cm wavelengths). This range is between the frequency of household microwave ovens (~2.45 GHz, 12.23 cm) and military radars (~10 GHz, 3 cm). At demonstrated frequencies, pulses of electromagnetic radiation moving through the material in one direction are composed of constituent waves moving in the opposite direction.

The metamaterial was constructed as a periodic array of copper split ring and wire conducting elements deposited onto a circuit-board substrate. The design was such that the cells, and the lattice spacing between the cells, were much smaller than the radiated electromagnetic wavelength. Hence, it behaves as an effective medium. The material has become notable because its range of (effective) permittivity $\varepsilon_{eff}$ and permeability $\mu_{eff}$ values have exceeded those found in any ordinary material. Furthermore, the characteristic of negative (effective) permeability evinced by this medium is particularly notable, because it has *not* been found in ordinary materials. In addition, the negative values for the magnetic component is directly related to its left-handed nomenclature, and properties (discussed in a section below).

The split-ring resonator (SRR), based on the prior 1999 theoretical article, is the tool employed to achieve negative permeability. This first composite *metamaterial* is then composed of split-ring resonators and electrical conducting posts.

Initially, these materials were only demonstrated at wavelengths longer than those in the visible spectrum. In addition, early NIMs were fabricated from opaque materials and usually made of non-magnetic constituents. As an illustration, however, if these materials are constructed at visible frequencies, and a flashlight is shone onto the resulting NIM slab, the material should focus the light at a point on the other side. This is not possible with a sheet of ordinary opaque material. In 2007, the NIST in collaboration with the Atwater Lab at Caltech created the first NIM active at optical frequencies. More recently (as of 2008), layered "fishnet" NIM materials made of silicon and silver wires have been integrated into optical fibres to create active optical elements.

### *Simultaneous Negative Permittivity and Permeability*

Negative permittivity $\varepsilon_{eff} < 0$ had already been discovered and realized in metals for frequencies all the way up to the plasma frequency, before the first metamaterial. There are two requirements to achieve a negative value for refraction. First, is to fabricate a material which can produce negative permeability $\mu_{eff} < 0$. Second, negative values for both permittivity and permeability must occur simultaneously over a common range of frequencies.

Therefore, for the first metamaterial, the nuts and bolts are one split-ring resonator electromagnetically combined with one (electric) conducting post. These are designed to resonate at designated frequencies to achieve the desired values. Looking at the make-up of the split ring, the associated magnetic field pattern from the SRR is dipolar. This dipolar behaviour is notable because this means it mimics nature's atom, but on a much larger scale, such as in this case at 2.5 millimetres. Atoms exist on the scale of picometres.

The splits in the rings create a dynamic where the SRR unit cell can be made resonant at radiated wavelengths *much larger* than the diametre of the rings. If the rings were closed, a half wavelength boundary would be electromagnetically imposed as a requirement for resonance.

The split in the second ring is oriented opposite to the split in the first ring. It is there to generate a large capacitance, which occurs in the small gap. This capacitance substantially decreases the resonant frequency while concentrating the electric field. The individual SRR depicted on the right had a resonant frequency of 4.845 GHz, and the resonance curve, inset in the graph, is also shown. The radiative losses from absorption and reflection are noted to be small, because the unit dimensions are much smaller than the free space, radiated wavelength.

When these units or cells are combined into a periodic arrangement, the magnetic coupling between the resonators is strengthened, and a *strong magnetic coupling occurs*. Properties unique in comparison to ordinary or conventional materials begin to emerge. For one thing, this periodic strong coupling creates a material, which now has an effective magnetic permeability $\mu_{eff}$ in response to the radiated-incident magnetic field.

## Composite Material Passband

Graphing the general dispersion curve, a region of propagation occurs from zero up to a lower band edge, followed by a gap, and then an upper passband. The presence of a 400 MHz gap between 4.2 GHz and 4.6 GHz implies a band of frequencies where $\mu_{eff} < 0$ occurs.

Furthermore, when wires are added symmetrically between the split rings, a passband occurs within the previously forbidden band of the split ring dispersion curves. That this passband occurs within a previously forbidden region indicates that the negative $\varepsilon_{eff}$ for this region has combined with the negative $\mu_{eff}$ to allow propagation, which fits with theoretical predictions. Mathematically, the dispersion relation leads to a band with negative group velocity everywhere, and a bandwidth that is independent of the plasma frequency, within the stated conditions.

Mathematical modelling and experiment have both shown that periodically arrayed conducting elements (non-magnetic by nature) respond predominately to the magnetic component of incident electromagnetic fields. The result is an effective medium and negative $\mu_{eff}$ over a band of frequencies. The permeability was verified to be the region of the forbidden band, where the gap in propagation occurred - from a finite section of material. This was combined with a negative permittivity material, $\varepsilon_{eff} < 0$, to form a "left-handed" medium, which formed a propagation band with negative group velocity where previously there was only attenuation. This validated predictions. In addition, a later work determined that this first metamaterial had a range of frequencies over which the refractive index was predicted to be negative for one direction of propagation. Other predicted electrodynamic effects were to be investigated in other research.

### Describing a Left-handed Material

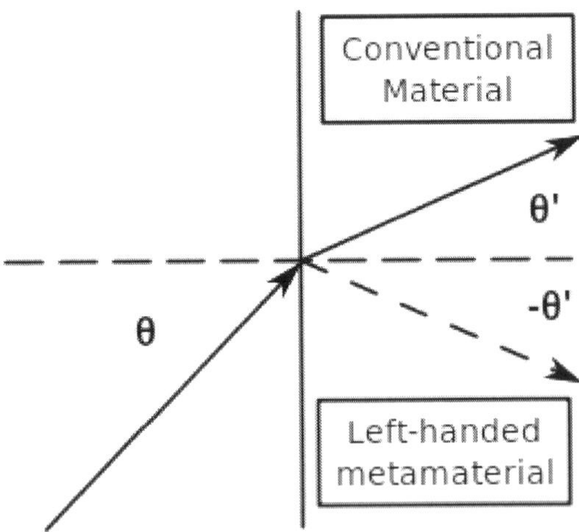


***Figure:*** *A comparison of refraction in a negative index metamaterial to that in a conventional material having the same, but positive refractive index. The incident beam θ enters from air and refracts in a normal (θ') or metamaterial (-θ').*

From the conclusions in the above section a left-handed material (LHM) can be defined. It is a material which exhibits simultancous negative values for permittivity, ε, and permeability, μ, in an overlapping frequency region. Since the values are derived from the effects of the composite medium system as a whole, these are defined as effective permittivity, $\varepsilon_{eff}$, and effective permeability, $\mu_{eff}$. Real values are then derived to denote the value of negative index of refraction, and wave vectors. This means that in practice losses will occur for a given medium used to transmit electromagnetic radiation such as microwave, or infrared frequencies, or visible light - for example. In this instance, real values describe either the amplitude or the intensity of a transmitted wave relative to an incident wave, while ignoring the negligible loss values.

### Isotropic Negative Index in Two Dimensions

In the above sections first fabricated metamaterial was constructed with resonating elements, which exhibited one direction of incidence and polarization. In other words, this structure exhibited left-handed propagation in one dimension. This was discussed in relation to Veselago's seminal work 33 years earlier (1967). He predicted that intrinsic to a material, which manifests negative values of effective permittivity and permeability, are several types of reversed physics phenomena. Hence, there was then a critical need for a higher dimensional LHMs to confirm Veselago's theory, as expected. The confirmation would include reversal of Snell's law (index of refraction), along with other reversed phenomena.

In the beginning of 2001 the existence of a higher dimensional structure was reported. It was two-dimensional and demonstrated by both experiment and numerical confirmation. It was an LHM, a composite constructed of wire strips mounted behind the split-ring resonators (SRRs) in a periodic configuration. It was created for the express purpose of being suitable for further experiments to produce the effects predicted by Veselago .

### *Experimental Verification of a Negative Index of Refraction*

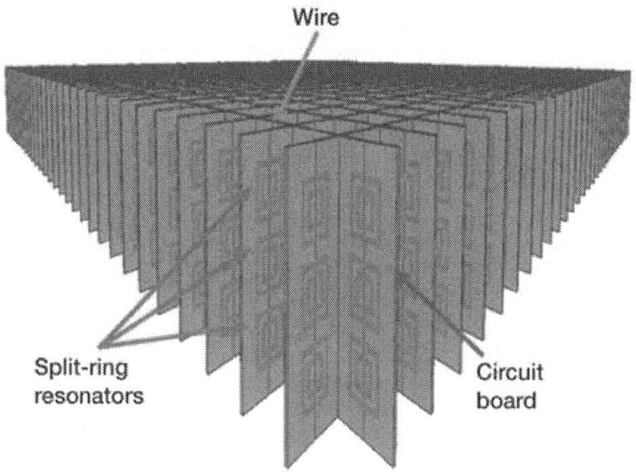


***Figure:*** *Split-ring resonator consisting of an inner square with a split on one side embedded in an outer square with a split on the other side. Split-ring resonators are on the front and right surfaces of the square grid, and single vertical wires are on the back and left surfaces.*

A theoretical work in published in 1967 by Soviet physicist Victor Veselago showed that a refractive index with negative values is possible and that this does not violate the laws of physics. As discussed previously (above), the first metamaterial had a range of frequencies over which the refractive index was predicted to be negative for one direction of propagation. It was reported in May 2000.

In 2001, a team of researchers constructed a prism composed of metamaterials (negative index metamaterials) to experimentally test for negative refractive index. The experiment used a waveguide to help transmit the proper frequency and isolate the material. This test achieved its goal because it successfully verified a negative index of refraction.

The experimental demonstration of negative refractive index was followed by another demonstration, in 2003, of a reversal of Snell's law, or reversed refraction. However, in this experiment negative index of refraction material is in free space from 12.6 to 13.2 GHz. Although the radiated frequency range is about the same, a notable distinction is this experiment is conducted in free space rather than employing waveguides.

Furthering the authenticity of negative refraction, the power flow of a wave transmitted through a dispersive left-handed material was calculated and compared to a dispersive right-handed material. The transmission of an incident field, composed of many frequencies, from an isotropic nondispersive material into an isotropic dispersive media is employed. The direction of power flow for both nondispersive and dispersive media is determined by the time-averaged Poynting vector. Negative refraction was shown to be possible for multiple frequency signals by explicit calculation of the Poynting vector in the LHM.

### *Fundamental Electromagnetic Properties of the NIM*

In a slab of conventional material with an ordinary refractive index – a right-handed material (RHM) – the wave front is transmitted away from the source. In a NIM the wavefront travels towards the source. However, the magnitude and direction of the flow of energy essentially remains the same in both the ordinary material and the NIM. Since the flow of energy remains the same in both materials (media), the impedance of the NIM matches the RHM. Hence, the sign of the intrinsic impedance is still positive in a NIM.

Light incident on a left-handed material, or NIM, will bend to the same side as the incident beam, and for Snell's law to hold, the refraction angle should be negative. In a passive metamaterial medium this determines a negative real and imaginary part of the refractive index.

### *Negative Refractive Index in Left-handed Materials*

In 1968 Victor Veselago's paper showed that the opposite directions of EM plane waves and the flow of energy was derived from the individual Maxwell curl equations. In ordinary optical materials, the curl equation for the electric field show a "right hand rule" for the directions of the electric field E, the magnetic induction B, and wave propagation, which goes in the direction of wave vector k. However, the direction of energy flow formed by E × H is right-handed only when *permeability is greater than zero*. This means that when permeability is less than zero, e.g. *negative*, wave propagation is reversed (determined by k), and contrary to the direction of energy flow. Furthermore, the relations of vectors E, H, and k form a *"left-handed" system* – and it was Veselago who coined the term "left-handed" (LH) material, which is in wide use today (2011). He contended that an LH material has a negative refractive index and relied on the steady-state solutions of Maxwell's equations as a centre for his argument.

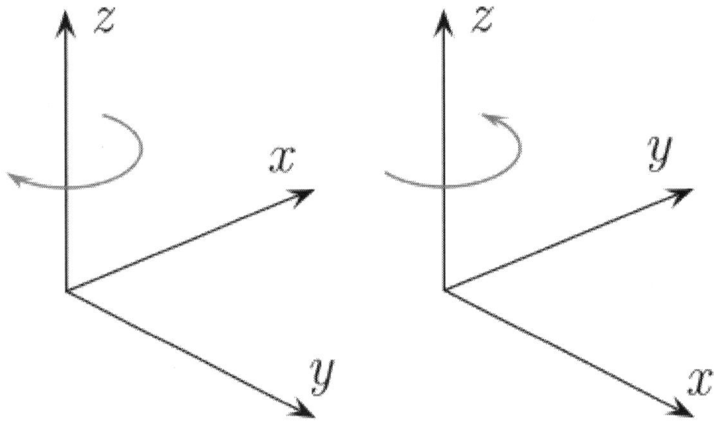


***Figure:*** *The left-handed orientation is shown on the left, and the right-handed on the right.*

After a 30-year void, when LH materials were finally demonstrated, it could be said that the designation of negative refractive index is unique to LH systems; even when compared to photonic crystals. Photonic crystals, like many other known systems, can exhibit unusual propagation behaviour such as reversal of phase and group velocities. But, negative refraction does not occur in these systems, and not yet realistically in Photonic crystals.

## *Negative Refraction at Visible Frequencies*

As of July 2013; in previous years, several anomalous studies have announced negative refraction at one single frequency or other in the visible spectrum. But the results of two such demonstrations are considered ambiguous by later studies. Another most recent, published, demonstration at one single visible frequency is still not the norm, or common, for the large body of work that has been produced in the field of metamaterials. To date, hundreds of scientific, peer reviewed, articles have been published regarding some aspect of metamaterials. In contrast, an insignificant number of studies have been published which claim apparent results in the visible spectrum. In an encyclopedia article such as this it is problematic to give undue weight to such studies, until these become common, or part of the norm, for metamaterials.

Moreover, although previous research efforts have announced negative refraction in the visible light spectrum (a single frequency), one April 2010 report stated that it was "the first one that operates on visible light." Also as before, the stated achievement is for one single frequency in the visible spectrum. In other words there is no broad band ability.

### *Experimental Verification of Reversed Cherenkov Radiation*

Besides reversed values for index of refraction, Veselago predicted the occurrence of reversed Cherenkov radiation (also known simply as CR) in a left-handed medium. In 1934 Pavel Cherenkov discovered a coherent radiation that occurs when certain types of media are bombarded by fast moving electron beams. In 1937 a theory built around CR stated that when charged particles, such as electrons, travel through a medium at speeds faster than the speed of light in the medium only then will CR radiate. As the CR occurs, electromagnetic radiation is emitted in a cone shape, fanning out in the forward direction.

CR and the 1937 theory has led to a large array of applications in high energy physics. A notable application are the Cherenkov counters. These are used to determine various properties of a charged particle such as its velocity, charge, direction of motion, and energy. These properties are important in the identification of different particles. For example, the counters were applied in the discovery of the antiproton and the J particle. Six large Cherenkov counters were used in the discovery of the J particle.

It has been difficult to experimentally prove the reversed Cherenkov radiation.

### *Other Optics with NIMs*

Theoretical work, along with numerical simulations, began early in the decade of the new 2000 millennium on the abilities of the DNG slab for subwavelength focusing. The research began with Pendry's proposed "Perfect lens." Several research investigations that followed

Pendry's concluded that the "Perfect lens" was possible in theory but impractical. One direction in subwavelength focusing proceeded with the use of negative index metamaterials, but based on the enhancements for imaging with surface plasmons. In another direction researchers explored paraxial approximations of NIM slabs.

### Implications of Negative Refractive Materials

The existence of negative refractive materials can result in a change in electrodynamic calculations for the case of *permeability* $\mu = 1$. A change from a conventional refractive index to a negative value gives incorrect results for conventional calculations, because some properties and effects have been altered. When *permeability* $\mu$ has values other than 1 this affects Snell's law, the Doppler effect, the Cherenkov radiation, Fresnel's equations, and Fermat's principle.

The refractive index is basic to the science of optics. Shifting the refractive index to a negative value may be a cause to revisit or reconsider the interpretation of some norms, or basic laws.

### US Patent on Left-handed Composite Media

The first US patent granted for a fabricated metamaterial is U.S. Patent 6,791,432, titled "Left handed composite media." The listed inventors are David R. Smith, Sheldon Schultz, Norman Kroll, Richard A. Shelby.

The invention achieves simultaneous negative permittivity and permeability over a common band of frequencies. The material can integrate media which is already composite or continuous, but which will produce negative permittivity and permeability within the same spectrum of frequencies. Different types of continuous or composite may be deemed appropriate when combined for the desired effect. However, the inclusion of a periodic array of conducting elements is preferred. The array scatters electromagnetic radiation at wavelengths longer than the size of the element and lattice spacing. The array is then viewed as an effective medium.

### Microscopic Explanation

At the microscale, an electromagnetic wave's phase speed is slowed in a material because the electric field creates a disturbance in the charges of each atom (primarily the electrons) proportional to the electric susceptibility of the medium. (Similarly, the magnetic field creates a disturbance proportional to the magnetic susceptibility.) As the electromagnetic fields oscillate in the wave, the charges in the material will be "shaken" back and forth at the same frequency. The charges thus radiate their own electromagnetic wave that is at the same frequency, but usually with a phase delay, as the charges may move out of phase with the force driving them. The light wave travelling in the medium is the macroscopic superposition (sum) of all such contributions in the material: the original wave plus the waves radiated by all the moving charges. This wave is typically a wave with the same frequency but shorter wavelength than the original, leading to a slowing of the wave's phase speed. Most of the radiation from oscillating material charges will modify the incoming wave, changing its velocity. However, some net energy will be radiated in other directions or even at other frequencies.

Depending on the relative phase of the original driving wave and the waves radiated by the charge motion, there are several possibilities:

- If the electrons emit a light wave which is 90° out of phase with the light wave shaking them, it will cause the total light wave to travel more slowly. This is the normal refraction of transparent materials like glass or water, and corresponds to a refractive index which is real and greater than 1.
- If the electrons emit a light wave which is 270° out of phase with the light wave shaking them, it will cause the total light wave to travel more quickly. This is called "anomalous refraction", and is observed close to absorption lines, with X-rays, and in some microwave systems. It corresponds to a refractive index less than 1. (Even though the phase velocity of light is greater than the speed of light in vacuum $c$, the signal velocity is not, as discussed above). If the response is sufficiently strong and out-of-phase, the result is negative refractive index discussed below.
- If the electrons emit a light wave which is 180° out of phase with the light wave shaking them, it will destructively interfere with the original light to reduce the total light intensity. This is light absorption in opaque materials and corresponds to an imaginary refractive index.
- If the electrons emit a light wave which is in phase with the light wave shaking them, it will amplify the light wave. This is rare, but occurs in lasers due to stimulated emission. It corresponds to an imaginary index of refraction, with the opposite sign as absorption.

For most materials at visible-light frequencies, the phase is somewhere between 90° and 180°, corresponding to a combination of both refraction and absorption.

## Dispersion (Optics)

In optics, dispersion is the phenomenon in which the phase velocity of a wave depends on its frequency, or alternatively when the group velocity depends on the frequency. Media having such a property are termed *dispersive media*. Dispersion is sometimes called *chromatic* dispersion to emphasize its wavelength-dependent nature, or group-velocity dispersion (GVD) to emphasize the role of the group velocity. Dispersion is most often described for light waves, but it may occur for any kind of wave that interacts with a medium or passes through an inhomogeneous geometry (e.g., a waveguide), such as sound waves. A material's dispersion is measured by its Abbe number, $V$, with low Abbe numbers corresponding to strong dispersion.

### *Examples of Dispersion*

The most familiar example of dispersion is probably a rainbow, in which dispersion causes the spatial separation of a white light into components of different wavelengths (different colours). However, dispersion also has an effect in many other circumstances: for example, GVD causes pulses to spread in optical fibres, degrading signals over long distances; also, a cancellation between group-velocity dispersion and nonlinear effects leads to soliton waves.

### *Sources of Dispersion*

There are generally two sources of dispersion: material dispersion and waveguide dispersion. Material dispersion comes from a frequency-dependent response of a material to waves. For example, material dispersion leads to undesired chromatic aberration in a lens or the separation of colours in a prism. Waveguide dispersion occurs when the speed of a wave in a waveguide (such as an optical fibre) depends on its frequency for geometric reasons, independent of any frequency dependence of the materials from which it is constructed. More generally, "waveguide" dispersion can occur for waves propagating through any inhomogeneous structure (e.g., a photonic crystal), whether or not the waves are confined to some region. In general, *both* types of dispersion may be present, although they are not strictly additive. Their combination leads to signal degradation in optical fibres for telecommunications, because the varying delay in arrival time between different components of a signal "smears out" the signal in time.

## Material Dispersion in Optics

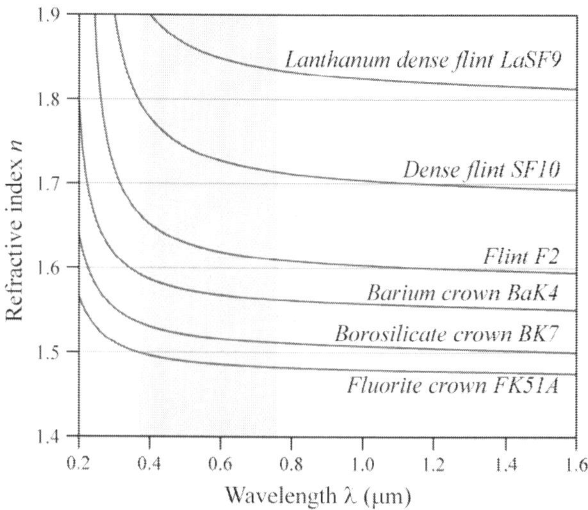


***Figure:*** *The variation of refractive index vs. vacuum wavelength for various glasses. The wavelengths of visible light are shaded in red.*

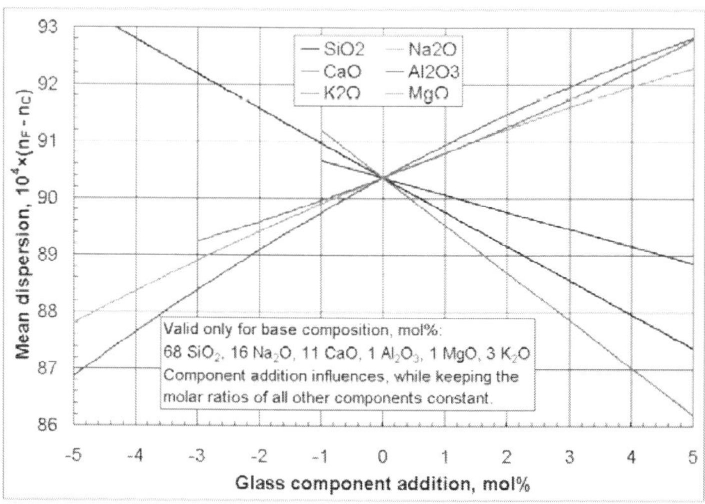


***Figure:*** *Influences of selected glass component additions on the mean dispersion of a specific base glass ($n_F$ valid for λ = 486 nm (blue), $n_C$ valid for λ = 656 nm (red))*

Material dispersion can be a desirable or undesirable effect in optical applications. The dispersion of light by glass prisms is used to construct spectrometres and spectroradiometres. Holographic gratings are also used, as they allow more accurate discrimination of wavelengths. However, in lenses, dispersion causes chromatic aberration, an undesired effect that may degrade images in microscopes, telescopes and photographic objectives.

The *phase velocity*, $v$, of a wave in a given uniform medium is given by

$$v = \frac{c}{n}$$

where $c$ is the speed of light in a vacuum and $n$ is the refractive index of the medium.

In general, the refractive index is some function of the frequency $f$ of the light, thus $n = n(f)$, or alternatively, with respect to the wave's wavelength $n = n(\lambda)$. The wavelength dependence of a material's refractive index is usually quantified by its Abbe number or its coefficients in an empirical formula such as the Cauchy or Sellmeier equations.

Because of the Kramers–Kronig relations, the wavelength dependence of the real part of the refractive index is related to the material absorption, described by the imaginary part of the refractive index (also called the extinction coefficient). In particular, for non-magnetic materials ($\mu = \mu_0$), the susceptibility $\chi$ that appears in the Kramers–Kronig relations is the electric susceptibility $\chi_e = n^2 - 1$.

The most commonly seen consequence of dispersion in optics is the separation of white light into a colour spectrum by a prism. From Snell's law it can be seen that the angle of refraction of light in a prism depends on the refractive index of the prism material. Since that refractive index varies with wavelength, it follows that the angle that the light is refracted by will also vary with wavelength, causing an angular separation of the colours known as *angular dispersion*.

For visible light, refraction indices $n$ of most transparent materials (e.g., air, glasses) decrease with increasing wavelength $\lambda$:

$$1 < n(\lambda_{\text{red}}) < n(\lambda_{\text{yellow}}) < n(\lambda_{\text{blue}}),$$

or alternatively:

$$\frac{dn}{d\lambda} < 0.$$

In this case, the medium is said to have *normal dispersion*. Whereas, if the index increases with increasing wavelength (which is typically the case for X-rays), the medium is said to have *anomalous dispersion*.

At the interface of such a material with air or vacuum, Snell's law predicts that light incident at an angle θ to the normal will be refracted at an angle arcsin(sin(θ)/*n*). Thus, blue light, with a higher refractive index, will be bent more strongly than red light, resulting in the well-known rainbow pattern.

## Group and Phase Velocity

Another consequence of dispersion manifests itself as a temporal effect. The formula $v = c / n$ calculates the *phase velocity* of a wave; this is the velocity at which the *phase* of any one frequency component of the wave will propagate. This is not the same as the *group velocity* of the wave, that is the rate at which changes in amplitude (known as the *envelope* of the wave) will propagate. For a homogeneous medium, the group velocity $v_g$ is related to the phase velocity by (here λ is the wavelength in vacuum, not in the medium):

$$v_g = c\left(n - \lambda \frac{dn}{d\lambda}\right)^{-1}.$$

The group velocity $v_g$ is often thought of as the velocity at which energy or information is conveyed along the wave. In most cases this is true, and the group velocity can be thought of as the *signal velocity* of the waveform. In some unusual circumstances, called cases of anomalous dispersion, the rate of change of the index of refraction with respect to the wavelength changes sign, in which case it is possible for the group velocity to exceed the speed of light ($v_g > c$). Anomalous dispersion occurs, for instance, where the wavelength of the light is close to an absorption resonance of the medium. When the dispersion is anomalous, however, group velocity is no longer an indicator of signal velocity. Instead, a signal travels at the speed of the wavefront, which is *c* irrespective of the index of refraction. Recently, it has become possible to create gases in which the group velocity is not only larger than the speed of light, but even negative. In these cases, a pulse can appear to exit a medium before it enters. Even in these cases, however, a signal travels at, or less than, the speed of light, as demonstrated by Stenner, et al.

The group velocity itself is usually a function of the wave's frequency. This results in group velocity dispersion (GVD), which causes a short pulse of light to spread in time as a result of different frequency components of the pulse travelling at different velocities. GVD is often quantified as the *group delay dispersion parametre* (again, this formula is for a uniform medium only):

$$D = -\frac{\lambda}{c}\frac{d^2n}{d\lambda^2}.$$

If $D$ is less than zero, the medium is said to have *positive dispersion*. If $D$ is greater than zero, the medium has *negative dispersion*. If a light pulse is propagated through a normally dispersive medium, the result is the higher frequency components travel slower than the lower frequency components. The pulse therefore becomes *positively chirped*, or *up-chirped*, increasing in frequency with time. Conversely, if a pulse travels through an anomalously dispersive medium, high frequency components travel faster than the lower ones, and the pulse becomes *negatively chirped*, or *down-chirped*, decreasing in frequency with time.

The result of GVD, whether negative or positive, is ultimately temporal spreading of the pulse. This makes dispersion management extremely important in optical communications systems based on optical fibre, since if dispersion is too high, a group of pulses representing a bit-stream will spread in time and merge, rendering the bit-stream unintelligible. This limits the length of fibre that a signal can be sent down without regeneration. One possible answer to this problem is to send signals down the optical fibre at a wavelength where the GVD is zero (e.g., around 1.3–1.5 μm in silica fibres), so pulses at this wavelength suffer minimal spreading from dispersion—in practice, however, this approach causes more problems than it solves because zero GVD unacceptably amplifies other nonlinear effects (such as four wave mixing). Another possible option is to use soliton pulses in the regime of anomalous dispersion, a form of optical pulse which uses a nonlinear optical effect to self-maintain its shape—solitons have the practical problem, however, that they require a certain power level to be maintained in the pulse for the nonlinear effect to be of the correct strength. Instead, the solution that is currently used in practice is to perform dispersion compensation, typically by matching the fibre with another fibre of opposite-sign dispersion so that the dispersion effects cancel; such compensation is ultimately limited by nonlinear effects such as self-phase modulation, which interact with dispersion to make it very difficult to undo.

Dispersion control is also important in lasers that produce short pulses. The overall dispersion of the optical resonator is a major factor in determining the duration of the pulses emitted by the laser. A pair of prisms can be arranged to produce net negative dispersion, which can be used to balance the usually positive dispersion of the laser medium. Diffraction gratings can also be used to produce dispersive effects; these are often used in high-power laser amplifier systems. Recently, an alternative to prisms and gratings has been developed: chirped mirrors. These dielectric mirrors are coated so that different wavelengths have different penetration lengths, and therefore different group delays. The coating layers can be tailored to achieve a net negative dispersion.

## Dispersion in Waveguides

Optical fibres, which are used in telecommunications, are among the most abundant types of waveguides. Dispersion in these fibres is one of the limiting factors that determine how much data can be transported on a single fibre.

The transverse modes for waves confined laterally within a waveguide generally have different speeds (and field patterns) depending upon their frequency (that is, on the relative size of the wave, the wavelength) compared to the size of the waveguide.

In general, for a waveguide mode with an angular frequency $\omega(\beta)$ at a propagation constant $\beta$ (so that the electromagnetic fields in the propagation direction *(z)* oscillate proportional to $e^{i(\beta z-\omega t)}$), the group-velocity dispersion parametre $D$ is defined as:

$$D=-\frac{2\pi c}{\lambda^2}\frac{d^2\beta}{d\omega^2}=\frac{2\pi c}{v_g^2\lambda^2}\frac{dv_g}{d\omega}$$

where $\lambda=2\pi c/\omega$ is the vacuum wavelength and $v_g=d\omega/d\beta$ is the group velocity. This formula generalizes the one in the previous section for homogeneous media, and includes both waveguide dispersion and material dispersion. The reason for defining the dispersion in this way is that $|D|$ is the (asymptotic) temporal pulse spreading $\Delta t$ per unit bandwidth $\Delta\lambda$ per unit distance travelled, commonly reported in ps / nm km for optical fibres.

A similar effect due to a somewhat different phenomenon is modal dispersion, caused by a waveguide having multiple modes at a given frequency, each with a different speed. A special case of this is polarization mode dispersion (PMD), which comes from a superposition of two modes that travel at different speeds due to random imperfections that break the symmetry of the waveguide.

### *Higher-order Dispersion over Broad Bandwidths*

When a broad range of frequencies (a broad bandwidth) is present in a single wavepacket, such as in an ultrashort pulse or a chirped pulse or other forms of spread spectrum transmission, it may not be accurate to approximate the dispersion by a constant over the entire bandwidth, and more complex calculations are required to compute effects such as pulse spreading.

In particular, the dispersion parametre *D* defined above is obtained from only one derivative of the group velocity. Higher derivatives are known as *higher-order dispersion.* These terms are simply a Taylor series expansion of the dispersion relation $\beta(\omega)$ of the medium or waveguide around some particular frequency. Their effects can be computed via numerical evaluation of Fourier transforms of the waveform, via integration of higher-order slowly varying envelope approximations, by a split-step method (which can use the exact dispersion relation rather than a Taylor series), or by direct simulation of the full Maxwell's equations rather than an approximate envelope equation.

## Dispersion in Gemology

### *Dispersion Values of Minerals*

In the technical terminology of gemology, *dispersion* is the difference in the refractive index of a material at the B and G (686.7 nm and 430.8 nm) or C and F (656.3 nm and 486.1 nm) Fraunhofer wavelengths, and is meant to express the degree to which a prism cut from the gemstone shows "fire", or colour. Dispersion is a material property. Fire depends on the dispersion, the cut angles, the lighting environment, the refractive index, and the viewer.

### *Dispersion in Imaging*

In photographic and microscopic lenses, dispersion causes chromatic aberration, which causes the different colours in the image not to overlap properly. Various techniques have been developed to counteract this, such as the use of achromats, multielement lenses with glasses of different dispersion. They are constructed in such a way that the chromatic aberrations of the different parts cancel out.

### *Dispersion in Pulsar Timing*

Pulsars are spinning neutron stars that emit pulses at very regular intervals ranging from milliseconds to seconds. Astronomers believe that the pulses are emitted simultaneously over a wide range of frequencies. However, as observed on Earth, the components of each pulse emitted at higher radio frequencies arrive before those emitted at lower frequencies. This dispersion occurs because of the ionized component of the interstellar medium, mainly the free electrons, which makes the group velocity frequency dependent. The extra delay added at a frequency $\nu$ is

$$t = k_{\mathrm{DM}} \times \left(\frac{\mathrm{DM}}{\nu^2}\right)$$

where the dispersion constant $k_{\mathrm{DM}}$ is given by

$$k_{\mathrm{DM}} = \frac{k_e e^2}{2\pi m_e c} \simeq 4.149 \mathrm{GHz}^2 \mathrm{pc}^{-1} \mathrm{cm}^3 \mathrm{ms},$$

and the dispersion measure *DM* is the column density of electrons — i.e. the number density of electrons $n_e$ (electrons/cm³) integrated along the path travelled by the photon from the pulsar to the Earth — and is given by

$$\mathrm{DM} = \int_0^d n_e \, dl$$

with units of parsecs per cubic centimetre ($1\mathrm{pc/cm}^3 = 30.857 \times 10^{21}\ \mathrm{m}^{-2}$).

Typically for astronomical observations, this delay cannot be measured directly, since the emission time is unknown. What *can* be measured is the difference in arrival times at two different frequencies. The delay $\Delta T$ between a high frequency $\nu_{hi}$ and a low frequency $\nu_{lo}$ component of a pulse will be

$$\Delta t = k_{\mathrm{DM}} \times \mathrm{DM} \times \left(\frac{1}{\nu_{\mathrm{lo}}^2} - \frac{1}{\nu_{\mathrm{hi}}^2}\right)$$

Re-writing the above equation in terms of *DM* allows one to determine the *DM* by measuring pulse arrival times at multiple frequencies. This in turn can be used to study the interstellar medium, as well as allow for observations of pulsars at different frequencies to be combined.

### *Complex Index of Refraction and Absorption*

When light passes through a medium, some part of it will always be absorbed. This can be conveniently taken into account by defining a complex index of refraction,

$$\tilde{n} = n + i\kappa.$$

Here, the real part of the refractive index *n* indicates the phase speed, while the imaginary part $\kappa$ indicates the amount of absorption loss when the electromagnetic wave propagates through the material.

That $\kappa$ corresponds to absorption can be seen by inserting this refractive index into the expression for electric field of a plane electromagnetic wave travelling in the $z$ -direction. We can do this by relating the wave number to the refractive index through $k = \frac{2\pi n}{\lambda_0}$ , with $\lambda_0$ being the vacuum wavelength. With complex wave number $\tilde{k}$ and refractive index $n + i\kappa$ this can be inserted into the plane wave expression as

$$E(z,t) = \text{Re}(E_0 e^{i(\tilde{k}z-\omega t)}) = \text{Re}(E_0 e^{i(2\pi(n+i\kappa)z/\lambda_0-\omega t)}) = e^{-2\pi\kappa z/\lambda_0}\text{Re}(E_0 e^{i(kz-\omega t)}).$$

Here we see that $\kappa$ gives an exponential decay, as expected from the Beer–Lambert law. Since intensity is proportional to the square of the electric field, the absorption coefficient becomes $\frac{4\pi\kappa}{\lambda_0}$ .

$\kappa$ is often called the extinction coefficient in physics although this has a different definition within chemistry. Both $n$ and $\kappa$ are dependent on the frequency. In most circumstances $\kappa > 0$ (light is absorbed) or $\kappa = 0$ (light travels forever without loss). In special situations, especially in the gain medium of lasers, it is also possible that $\kappa < 0$, corresponding to an amplification of the light.

An alternative convention uses $\tilde{n} = n - i\kappa$ instead of $\tilde{n} = n + i\kappa$ , but where $\kappa > 0$ still corresponds to loss. Therefore these two conventions are inconsistent and should not be confused. The difference is related to defining sinusoidal time dependence as $\text{Re}(e^{-i\omega t})$ versus $\text{Re}(e^{+i\omega t})$.

Dielectric loss and non-zero DC conductivity in materials cause absorption. Good dielectric materials such as glass have extremely low DC conductivity, and at low frequencies the dielectric loss is also negligible, resulting in almost no absorption (κ H" 0). However, at higher frequencies (such as visible light), dielectric loss may increase absorption significantly, reducing the material's transparency to these frequencies.

The real and imaginary parts of the complex refractive index are related through the Kramers–Kronig relations. For example, one can determine a material's full complex refractive index as a function of wavelength from an absorption spectrum of the material.

For X-ray and extreme ultraviolet radiation the complex refractive index deviates only slightly from unity and usually has a real part smaller than 1. It is therefore normally written as $\tilde{n} = 1 - \delta + i\beta$ (or ) $\tilde{n} = 1 - \delta - i\beta$ .

### *Relations to other Quantities*

***Phase Speed:*** The refractive index is the ratio of the speed of light in vacuum and the phase speed of light in a material,

$$n = \frac{c}{v_{\text{phase}}}$$

The phase speed is defined as the rate at which the crests of the waveform propagate; that is, the rate at which the phase of the waveform is moving. The *group speed* is the rate at which the *envelope* of the waveform is propagating; that is, the rate of variation of the amplitude of the waveform. Provided the waveform is not distorted significantly during propagation, it is the group speed that represents the rate at which information (and energy) may be transmitted by the wave (for example, the speed at which a pulse of light travels down an optical fibre). For the analytic properties constraining the unequal phase and group speeds in dispersive media, refer to dispersion (optics).

### Refraction

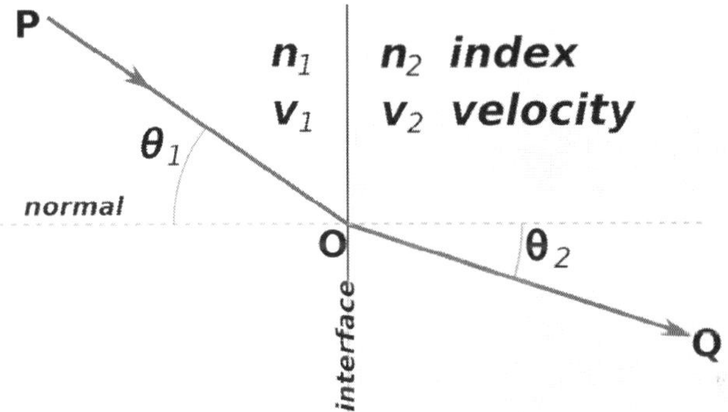


***Figure:*** *Refraction of light at the interface between two media of different refractive indices, with* $n_2 > n_1$. *Since the phase velocity is lower in the second medium* ($v_2 < v_1$)*, the angle of refraction* $\theta_2$ *is less than the angle of incidence* $\theta_1$*; that is, the ray in the higher-index medium is closer to the normal.*

When light moves from one medium to another as in the figure to the right, it changes direction, i.e. it is refracted. If it goes from a medium with refractive index $n_1$ to one with refractive index $n_2$, with an incidence angle to the surface normal of $\theta_1$, the transmission angle $\theta_2$ can be calculated from Snell's law:

$$n_1 \sin\theta_1 = n_2 \sin\theta_2 .$$

If there is no angle $\theta_2$ fulfilling Snell's law, i.e.,

$$\frac{n_1}{n_2}\sin\theta_1 > 1,$$

the light cannot be transmitted and will instead undergo total internal reflection.

### *Reflectivity*

Apart from the transmitted light there is also a reflected part. The reflection angle is equal to the incidence angle, and the amount of light that is reflected is determined by the reflectivity of the surface. The reflectivity can be calculated from the refractive index and the incidence angle with the Fresnel equations, which for normal incidence reduces to

$$R = \left| \frac{n_1 - n_2}{n_1 + n_2} \right|^2 .$$

For common glass in air, $n_1 = 1$ and $n_2 = 1.5$, and thus about 4% of the incident power is reflected. At other incidence angles the reflectivity will also depend on the polarization of the incoming light. At a certain angle called Brewster's angle, p-polarized light (light with the electric field in the plane of incidence) will be totally transmitted. Brewster's angle can be calculated from the two refractive indices of the interface as

$$\theta_B = \arctan\left(\frac{n_2}{n_1}\right).$$

### *Lenses*

The focal length of a lens is determined by its refractive index $n$ and the radii of curvature $R_1$ and $R_2$ of its surfaces. The power of a thin lens in air is given by the Lensmaker's formula:

$$\frac{1}{f} = (n-1)\left(\frac{1}{R_1} - \frac{1}{R_2}\right).$$

### *Dielectric Constant*

The refractive index of electromagnetic radiation equals

$$n = \sqrt{\epsilon_r \mu_r},$$

where $\epsilon_r$ is the material's relative permittivity, and $\mu_r$ is its relative permeability. For most naturally occurring materials, $\mu_r$ is very close to 1 at optical frequencies, therefore $n$ is approximately $\sqrt{\epsilon_r}$.

The frequency dependent dielectric constant is simply the square of the (complex) refractive index in a non-magnetic medium (one with a relative permeability of unity). The refractive index is used for optics in Fresnel equations and Snell's law; while the dielectric constant is used in Maxwell's equations and electronics.

Where $\tilde{\epsilon}$ is the complex dielectric constant with real and imaginary parts $\epsilon_1$ and $\epsilon_2$, and $n$ and $k$ are the real and imaginary parts of the refractive index, all functions of frequency:

$$\tilde{\epsilon} = \epsilon_1 + i\epsilon_2 = (n + i\kappa)^2.$$

Conversion between refractive index and dielectric constant is done by:

$$\epsilon_1 = n^2 - \kappa^2$$

$$\epsilon_2 = 2n\kappa$$

$$n = \sqrt{\frac{\sqrt{\epsilon_1^2 + \epsilon_2^2} + \epsilon_1}{2}}$$

$$\kappa = \sqrt{\frac{\sqrt{\epsilon_1^2 + \epsilon_2^2} - \epsilon_1}{2}}.$$

## Density

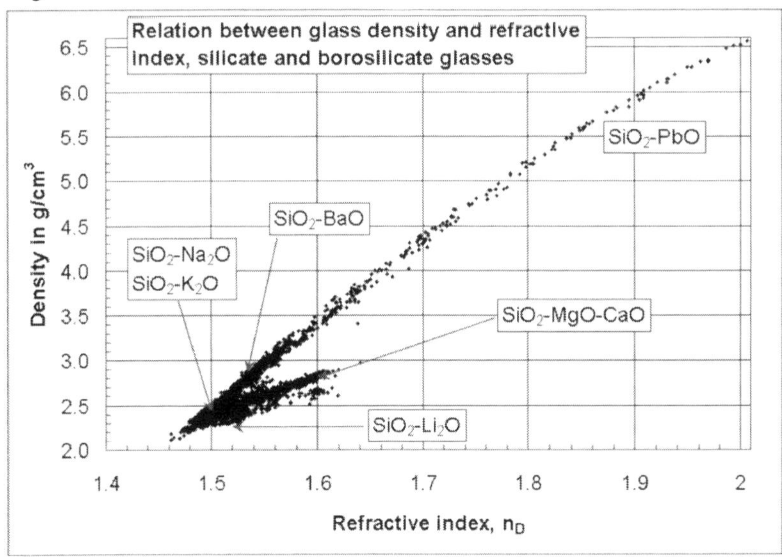


***Figure:*** *Relation between the refractive index and the density of silicate and borosilicate glasses.*

In general, the refractive index of a glass increases with its density. However, there does not exist an overall linear relation between the refractive index and the density for all silicate and borosilicate glasses. A relatively high refractive index and low density can be obtained with glasses containing light metal oxides such as $Li_2O$ and MgO, while the opposite trend is observed with glasses containing PbO and BaO as seen in the diagram at the right.

Many oils (such as olive oil) and ethyl alcohol are examples of liquids which are more refractive, but less dense, than water, contrary to the general correlation between density and refractive index.

## *Group Index*

Sometimes, a "group speed refractive index", usually called the *group index* is defined:

$$n_g = \frac{c}{v_g}$$

where $v_g$ is the group velocity. This value should not be confused with $n$, which is always defined with respect to the phase velocity. When the dispersion is small, the group velocity can be linked to the phase velocity by the relation

$$v_g = v - \lambda \frac{dv}{d\lambda}.$$

In this case the group index can thus be written in terms of the wavelength dependence of the refractive index as

$$n_g = \frac{n}{1 + \frac{\lambda}{n}\frac{dn}{d\lambda}},$$

where $\lambda$ is the wavelength in the medium.

When the refractive index of a medium is known as a function of the vacuum wavelength (instead of the wavelength in the medium), the corresponding expressions for the group velocity and index are (for all values of dispersion)

$$v_g = c\left(n - \lambda_0 \frac{dn}{d\lambda_0}\right)^{-1}.$$

$$n_g = n - \lambda_0 \frac{dn}{d\lambda_0},$$

where $\lambda_0$ is the wavelength in vacuum.

### *Momentum (Abraham-Minkowski Controversy)*

In 1908, Hermann Minkowski calculated the momentum of a refracted ray, $p$, where $E$ is energy of the photon, $c$ is the speed of light in vacuum and $n$ is the refractive index of the medium as follows:

$$p = \frac{nE}{c}.$$

In 1909, Max Abraham proposed the following formula for this calculation:

$$p = \frac{E}{nc}.$$

A 2010 study suggested that *both* equations are correct, with the Abraham version being the kinetic momentum and the Minkowski version being the canonical momentum, and claims to explain the contradicting experimental results using this interpretation.

### *Other Relations*

As shown in the Fizeau experiment, when light is transmitted through a moving medium, its speed relative to a stationary observer is:

$$V = \frac{c}{n} + v\left(1 - \frac{1}{n^2}\right)$$

The refractive index of a substance can be related to its polarizability with the Lorentz–Lorenz equation or to the molar refractivities of its constituents by the Gladstone–Dale relation.

### *Refractivity*

In atmospheric applications, the refractivity is defined as $N = 10^6(n - 1)$. The $10^6$ factor is chosen because for air, $n$ deviates from unity at most a few parts per ten thousand.

### *Nonscalar, Nonlinear, or Nonhomogeneous Refraction*

So far, we have assumed that refraction is given by linear equations involving a spatially constant, scalar refractive index. These assumptions can break down in different ways, to be described in the following subsections.

## Birefringence

Birefringence is the optical property of a material having a refractive index that depends on the polarization and propagation direction of light. These optically anisotropic materials are said to be birefringent (or birefractive). The birefringence is often quantified as the maximum difference between refractive indices exhibited by the material. Crystals with asymmetric crystal structures are often birefringent, as well as plastics under mechanical stress.

Birefringence is responsible for the phenomenon of double refraction whereby a ray of light, when incident upon a birefringent material, is split by polarization into two rays taking slightly different paths. This effect was first described by the Danish scientist Rasmus Bartholin in 1669, who observed it in calcite, a crystal having one of the strongest birefringences. However it wasn't until the 19th century that Augustin-Jean Fresnel correctly described the phenomenon in terms of polarization, understanding light as a wave with field components in transverse polarizations (perpendicular to the direction of the wave vector).

### *Explanation*

The simplest (and most common) type of birefringence is that of materials with uniaxial anisotropy. That is, the structure of the material is such that it has an axis of symmetry with all perpendicular directions optically equivalent. This axis is known as the optic axis of the material, and components of light with linear polarizations parallel and perpendicular to it have unequal indices of refraction, denoted $n_e$ and $n_o$, respectively, where the subscripts stand for *extraordinary* and *ordinary*. The names reflect the fact that, if unpolarized light enters the material at some angle of incidence, the component of the incident radiation whose polarization is perpendicular to the optic axis will be refracted according to the standard law of refraction for a material of refractive index $n_o$, while the other polarization component, the so-called *extraordinary ray* will refract at a different angle determined by the angle of incidence, the orientation of the optic axis, and the birefringence

$$\Delta n = n_e - n_o.$$

What's more, the extraordinary ray is an inhomogeneous wave whose power flow (given by the Poynting vector) is not exactly parallel to the wave vector. This causes a shift in that beam, even when launched at normal incidence, that is popularly observed using a crystal of calcite as photographed above. Rotating the calcite crystal will cause one of the two images, that of the extraordinary ray, to rotate slightly around that of the ordinary ray which remains fixed.

When the light propagates either along or orthogonal to the optic axis, such a lateral shift does not occur. In the first case, both polarizations see the same effective refractive index, so there is no extraordinary ray. In the second case the extraordinary ray propagates at a different phase velocity (corresponding to $n_e$) but is not an inhomogeneous wave. A crystal with its optic axis in this orientation, parallel to the optical surface, may be used to create a waveplate, in which there is no distortion of the image but an intentional modification of the state of polarization of the incident wave. For instance, a quarter-wave plate is commonly used to create circular polarization from a linearly polarized source.

The more general case of biaxially anisotropic materials, also known as trirefringent materials, is substantially more complex. Then there are *three* refractive indices corresponding to three principal axes of the crystal. Generally *both* polarizations are extraordinary rays with different effective refractive indices which can be determined using the index ellipsoid for a given polarization vector.

### Sources of Optical Birefringence

While birefringence is usually obtained using an anisotropic crystal, it can result from an optically isotropic material in a few ways:

- *Stress birefringence* results when isotropic materials are stressed or deformed such that the isotropy is lost in one direction (i.e., stretched or bent). Example
- By the Kerr effect, whereby an applied electric field induces birefringence at optical frequencies through the effect of nonlinear optics;
- By the Faraday effect, where a magnetic field causes some materials to become *circularly birefringent* (having slightly different indices of refraction for left and right handed circular polarizations), making the material *optically active* until the field is removed;
- By self or forced alignment of highly polar molecules such as lipids, some surfactants or liquid crystals, that generate highly birefringent thin films.

### Examples of Uniaxial Birefringent Materials

***Table:*** *Uniaxial materials, at 590 nm*

| ***Material*** | ***Crystal system*** | $n_o$ | $n_e$ | $\Delta n$ |
|---|---|---|---|---|
| barium borate $BaB_2O_4$ | Trigonal | 1.6776 | 1.5534 | -0.1242 |
| beryl $Be_3Al_2(SiO_3)_6$ | Hexagonal | 1.602 | 1.557 | -0.045 |

*Contd...*

| ***Material*** | ***Crystal system*** | $n_o$ | $n_e$ | $\Delta n$ |
|---|---|---|---|---|
| calcite $CaCO_3$ | Trigonal | 1.658 | 1.486 | -0.172 |
| ice $H_2O$ | Hexagonal | 1.309 | 1.313 | +0.004 |
| lithium niobate $LiNbO_3$ | Trigonal | 2.272 | 2.187 | -0.085 |
| magnesium fluoride $MgF_2$ | Tetragonal | 1.380 | 1.385 | +0.006 |
| quartz $SiO_2$ | Trigonal | 1.544 | 1.553 | +0.009 |
| ruby $Al_2O_3$ | Trigonal | 1.770 | 1.762 | -0.008 |
| rutile $TiO_2$ | Tetragonal | 2.616 | 2.903 | +0.287 |
| sapphire $Al_2O_3$ | Trigonal | 1.768 | 1.760 | -0.008 |
| silicon carbide SiC | Hexagonal | 2.647 | 2.693 | +0.046 |
| tourmaline (complex silicate ) | Trigonal | 1.669 | 1.638 | -0.031 |
| zircon, high $ZrSiO_4$ | Tetragonal | 1.960 | 2.015 | +0.055 |
| zircon, low $ZrSiO_4$ | Tetragonal | 1.920 | 1.967 | +0.047 |

The best-studied birefringent materials are crystalline; the refractive indices (at wavelength ~ 590 nm) of several such uniaxial crystals are tabulated to the right .

Many plastics are birefringent, because their molecules are 'frozen' in a stretched conformation when the plastic is molded or extruded. For example, ordinary cellophane is birefringent. Polarizers are routinely used to detect stress in plastics such as polystyrene and polycarbonate.

Cotton (gossypium hirsutum) fibre is birefringent because of high levels of cellulosic material in the fibre's secondary cell wall.

Inevitable manufacturing imperfections in optical fibre leads to birefringence which is one cause of pulse broadening in fibre-optic communications. Such imperfections can be geometrical (lack of circular symmetry), due to stress applied to the optical fibre, and/or due to bending of the fibre. Birefringence is *intentionally* introduced (for instance, by making the cross-section elliptical) in order to produce polarization-maintaining optical fibres.

In addition to anisotropy in the electric polarizability (electric susceptibility), anisotropy in the magnetic polarizability magnetic permeability]) will also cause birefringence. However at optical frequencies, values of magnetic permeability for natural materials are not measurably different from $\mu_0$ so this is not a source of optical birefringence in practice.

### *Fast and Slow Rays*

| ***Effective refractive indices in negative uniaxial materials*** | | | | |
|---|---|---|---|---|
| ***Propagation direction*** | ***Ordinary ray*** | | ***Extraordinary ray*** | |
| | ***Polarization*** | $n_{eff}$ | ***Polarization*** | $n_{eff}$ |
| $z$ | $xy$-plane | $n_o$ | n/a | n/a |
| $xy$-plane | $xy$-plane | $n_o$ | $z$ | $n_e$ |
| $xz$-plane | $y$ | $n_o$ | $xz$-plane | $n_e < n < n_o$ |
| other | analogous to $xz$-plane | | | |

For a given propagation direction, in general there are two perpendicular polarizations for which the medium behaves as if it had a single effective refractive index. In a uniaxial material, these polarizations are called the extraordinary and the ordinary ray (*e* and *o* rays), with the ordinary ray having the effective refractive index $n_o$. A biaxial crystal is characterized by three refractive indices $\alpha$, $\beta$, and $\gamma$ applying to its principle axes. A wave in a specified direction will consist of two polarization components with generally different effective refractive indices. The so-called slow ray is the component for which the material has the higher effective refractive index (slower phase velocity), while the fast ray has a lower effective refractive index.

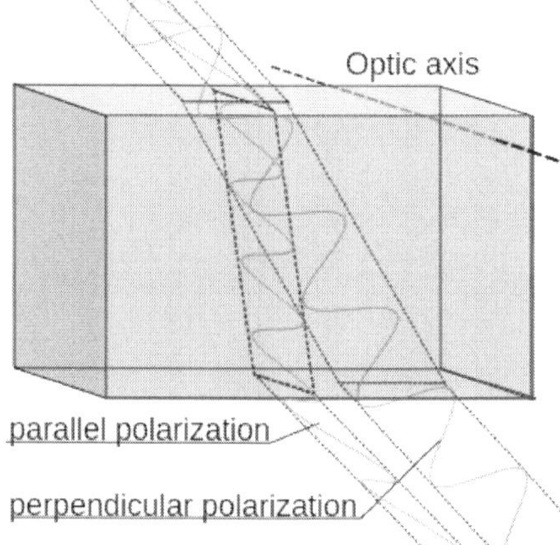


***Figure:*** *Rays passing through a positively birefringent material. The incident light has parallel and perpendicular polarisation components (linear polarization at 45° the optic axis). The optic axis is perpendicular to the direction of the perpendicular component of incident ray, so the ray polarized parallel to the optic axis has a greater refractive index than the ray polarized perpendicular to it.*

For a uniaxial material with the $z$ axis defined to be the optical axis, the effective refractive indices are as in the table on the right. For rays propagating in directions other than $z$, the effective refractive index of the extraordinary ray is in between $n_o$ and $n_e$, depending on the angle between the polarization vector and the $z$ axis. The effective refractive index can be determined using the index ellipsoid.

## Positive or Negative

Uniaxial birefringence is classified as positive when the extraordinary index of refraction $n_e$ is greater than the ordinary index $n_o$. Negative birefringence means that $\Delta n = n_e - n_o$ is less than zero. In other words, the polarization of the fast (or slow) wave is perpendicular to the optical axis when the birefringence of the crystal is positive (or negative, respectively). The terms "positive" and "negative" are not applied in the case of biaxial crystals, since all three of the principal axes have different refractive indices, rather than two being the same but different fmrom the one that's designated as the optic axis in a uniaxial crystal.

***Table:*** *Biaxial Materials, at 590 nm*

| *Material* | *Crystal system* | $n_\alpha$ | $n_\beta$ | $n_\gamma$ |
|---|---|---|---|---|
| borax $Na_2(B_4O_5)(OH)_4 \cdot 8(H_2O)$ | Monoclinic | 1.447 | 1.469 | 1.472 |
| epsom salt $MgSO_4 \cdot 7(H_2O)$ | Monoclinic | 1.433 | 1.455 | 1.461 |
| mica, biotite K(Mg,Fe)3AlSi3O10(F,OH)2 | Monoclinic | 1.595 | 1.640 | 1.640 |
| mica, muscovite $KAl_2(AlSi_3O_{10})(F,OH)_2$ | Monoclinic | 1.563 | 1.596 | 1.601 |
| olivine $(Mg, Fe)_2SiO_4$ | Orthorhombic | 1.640 | 1.660 | 1.680 |
| perovskite $CaTiO_3$ | Orthorhombic | 2.300 | 2.340 | 2.380 |
| topaz $Al_2SiO_4(F,OH)_2$ | Orthorhombic | 1.618 | 1.620 | 1.627 |
| ulexite $NaCaB_5O_6(OH)_6 \cdot 5(H_2O)$ | Triclinic | 1.490 | 1.510 | 1.520 |

### *Biaxial Birefringence*

Biaxial birefringence, also known as trirefringence, describes an anisotropic material in which the optical properties are not invariant under rotation about a particular axis (the optic axis, in uniaxial crystals). For such a material, the refractive index tensor n, will in general have three distinct eigenvalues that can be labelled $n_\alpha$, $n_\beta$ and $n_\gamma$.

### *Measurement*

Birefringence and other polarization based optical effects (such as optical rotation and linear or circular dichroism) can be measured by measuring the changes in the polarization of light passing through the material. These measurements are known as polarimetry. Birefringence of lipid bilayers can be measured using dual polarisation interferometry. This provides a measure of the degree of order within these fluid layers and how this order is disrupted when the layer interacts with other biomolecules.

### Applications

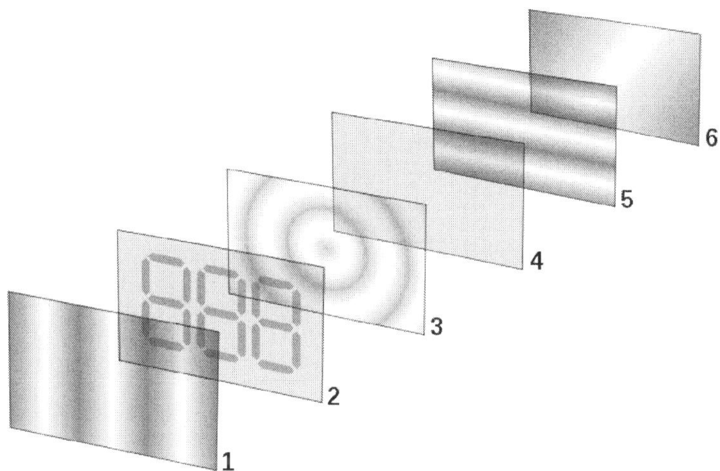


***Figure:*** *Reflective twisted nematic liquid crystal display. Light reflected by surface (6) (or coming from a backlight) is horizontally polarized (5) and passes through the liquid crystal modulator (3) sandwiched in between transparent layers (2,4) containing electrodes. Horizontally polarized light is blocked by the vertically oriented polarizer (1) except where its polarization has been rotated by the liquid crystal (3), appearing bright to the viewer*

Birefringence is used in many optical devices. Liquid crystal displays, the most common sort of flat panel display, cause their pixels to become lighter or darker through rotation of the polarization (circular birefringence) of linearly polarized light as viewed through a sheet polarizer at the screen's surface. Similarly, light modulators modulate the intensity of light through electrically induced birefringence of polarized light followed by a polarizer. The Lyot filter is a specialized narrowband spectral filter employing the wavelength dependence of birefringence. Wave plates are thin birefringent sheets widely used in certain optical equipment for modifying the polarization state of light passing through it.

Birefringence also plays an important role in second harmonic generation and other nonlinear optical components, as the crystals used for this purpose are almost always birefringent. By adjusting the angle of incidence, the effective refractive index of the extraordinary ray can be tuned in order to achieve phase matching which is required for efficient operation of these devices.

### Medicine

Birefringence is utilized in medical diagnostics. One powerful accessory used with optical microscopes is a pair of crossed polarizing filters. Light from the source is polarized in the X direction after passing through the first polarizer, but above the specimen is a polarizer (a so-called *analyzer*) oriented in the Y direction. Therefore no light from the source will be accepted by the analyzer, and the field will appear dark. However areas of the sample possessing birefringence will generally couple some of the X polarized light into the Y polarization; these areas will then appear bright against the dark background. Modifications to this basic principle can differentiate between positive and negative birefringence.

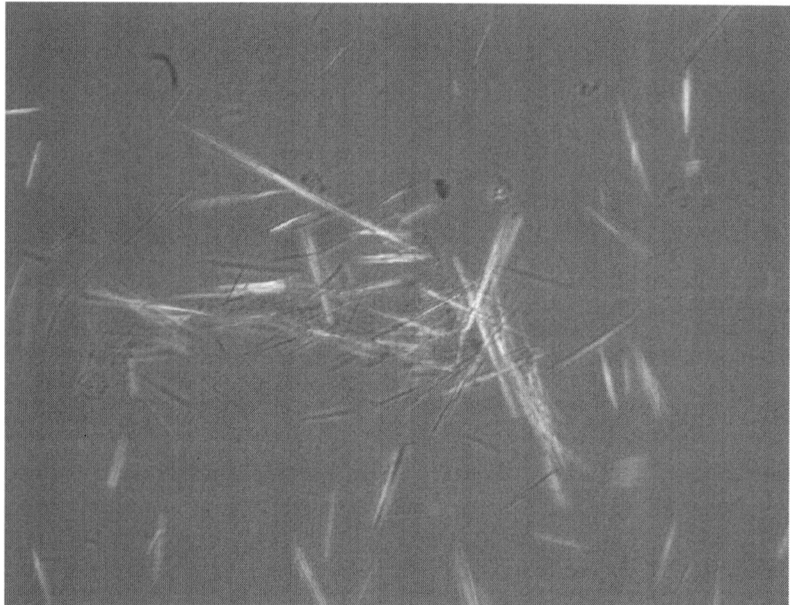

***Figure:*** *Urate crystals, with the crystals with their long axis seen as horizontal in this view being parallel to that of a red compensator filter. These appear as yellow, and are thereby of negative birefringence.*

For instance, needle aspiration of fluid from a gouty joint will reveal negatively birefringent monosodium urate crystals. Calcium pyrophosphate crystals, in contrast, show weak positive birefringence. Urate crystals appear yellow and calcium pyrophosphate crystals appear blue when their long axes are aligned parallel to that of a red compensator filter, or a crystal of known birefringence is added to the sample for comparison.

Birefringence can be observed in amyloid plaques such as are found in the brains of Alzheimer's patients when stained with a dye such as Congo Red. Modified proteins such as immunoglobulin light chains abnormally accumulate between cells, forming fibrils. Multiple folds of these fibres line up and take on a beta-pleated sheet conformation. Congo red dye intercalates between the folds and, when observed under polarized light, causes birefringence.

In ophthalmology, scanning laser polarimetry utilises the birefringence of the optic nerve fibre layer to indirectly quantify its thickness, which is of use in the assessment and monitoring of glaucoma.

Birefringence characteristics in sperm heads allow for the selection of spermatozoa for intracytoplasmic sperm injection. Likewise, *zona imaging* uses birefringence on oocytes to select the ones with highest chances of successful pregnancy. Birefringence of particles biopsied from pulmonary nodules indicates silicosis.

### Stress Induced Birefringence

**Figure:** *Colour pattern of a plastic box with "frozen in" mechanical stress placed between two crossed polarizers.*

Isotropic solids do not exhibit birefringence. However, when they are under mechanical stress, birefringence results. The stress can be applied externally or is 'frozen' in after a birefringent plastic ware is cooled after it is manufactured using injection molding. When such a sample is placed between two crossed polarizers, colour patterns can be observed, because polarization of a light ray is rotated after passing through a birefingent material and the amount of rotation is dependent on wavelength. The experimental method called photoelasticity used for analyzing stress distribution in solids is based on the same principle.

### Other Cases of Birefringence

Birefringence is observed in anisotropic elastic materials. In these materials, the two polarizations split according to their effective refractive indices which are also sensitive to stress. The study of birefringence in shear waves travelling through the solid earth (the earth's liquid core doesn't support shear waves) is widely used in seismology. Birefringence is widely used in mineralogy to identify rocks, minerals, and gemstones.

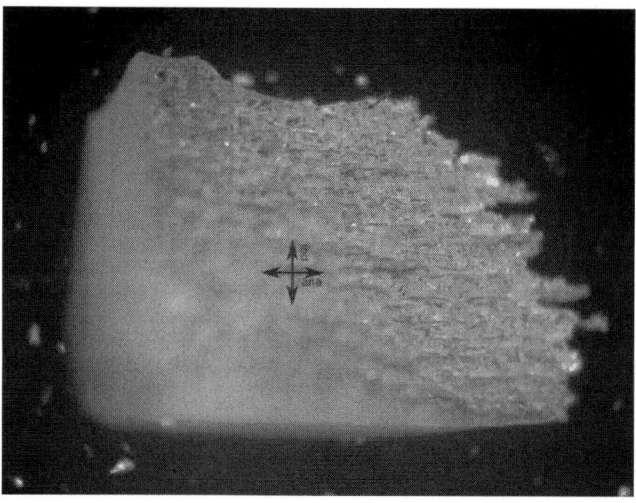

***Figure:*** *Birefringent rutile observed in different polarizations using a rotating polarizer (or* analyzer*)*

### Theory

Birefringence results when a material's permittivity is not describable using a scalar value, but requires a tensor to relate the electric displacement (*D*) with the electric field (*E*). Consider a plane wave propagating in an anisotropic medium, with a permittivity tensor ε and assuming no magnetic permeability in the medium: $\mu = \mu_0$. We shall assume that the electric field of a wave of angular frequency ω can be written in the form:

$$\mathrm{E} = \mathrm{E}_0 \exp\left[i(\mathrm{k}\cdot\mathrm{r} - \omega t)\right] \tag{2}$$

where r is the position vector, $t$ is time, and $\mathrm{E}_0$ is a vector describing the electric field at r=0, $t$=0. Then we shall find the possible wave vectors k using Maxwell's equations from which we obtain:

$$-\nabla \times \nabla \times \mathrm{E} = \mu_0 \frac{\partial^2 \mathrm{D}}{\partial t^2} \tag{3a}$$

$$\nabla\cdot\mathrm{D} = 0 \tag{3b}$$

where the so-called electric displacement vector D is now related to the electric field E through the permittivity tensor ε:

$$\mathrm{D} = \epsilon \mathrm{E}.$$

Substituting the definition of D and eqn. 2 into eqns. 3a-b leads to the conditions:

$$|\mathrm{k}|^2\, \mathrm{E}_0 - (\mathrm{k}\cdot\mathrm{E}_0)\mathrm{k} = \mu_0\omega^2(\epsilon \mathrm{E}_0) \tag{4a}$$

$$\mathrm{k}\cdot(\epsilon \mathrm{E}_0) = 0 \tag{4b}$$

Eqn. 4b indicates that D is orthogonal to the direction of the wavevector k, even though that is no longer generally true for E as would be the case in an isotropic medium.

To find the allowed values of k, $\mathrm{E}_0$ can be eliminated from eq 4a. If eqn 4a is written in Cartesian coordinates with the $x$, $y$ and $z$ axes chosen in the principle directions of the permitivity tensor ε, then

$$\epsilon = \epsilon_0 \begin{bmatrix} n_x^2 & 0 & 0 \\ 0 & n_y^2 & 0 \\ 0 & 0 & n_z^2 \end{bmatrix} \tag{4c}$$

where the diagonal values are the refractive indices for polarizations along the three principal axes $x$, $y$ and $z$. With ε in this form, and noting that the speed of light $c = 1/\sqrt{\mu_0\epsilon_0}$, eqn. 4a becomes

$$\left(-k_y^2 - k_z^2 + \frac{\omega^2 n_x^2}{c^2}\right) E_x + k_x k_y E_y + k_x k_z E_z = 0 \tag{5a}$$

$$k_x k_y E_x + \left(-k_x^2 - k_z^2 + \frac{\omega^2 n_y^2}{c^2}\right) E_y + k_y k_z E_z = 0 \tag{5b}$$

$$k_x k_z E_x + k_y k_z E_y + \left(-k_x^2 - k_y^2 + \frac{\omega^2 n_z^2}{c^2}\right) E_z = 0 \tag{5c}$$

where $E_x$, $E_y$, $E_z$, $k_x$, $k_y$ and $k_z$ are the components of $E_0$ and k. This is a set of linear equations in $E_x$, $E_y$, $E_z$, and they have a non-trivial solution if the following determinant is zero:

$$\begin{vmatrix} \left(-k_y^2 - k_z^2 + \frac{\omega^2 n_x^2}{c^2}\right) & k_x k_y & k_x k_z \\ k_x k_y & \left(-k_x^2 - k_z^2 + \frac{\omega^2 n_y^2}{c^2}\right) & k_y k_z \\ k_x k_z & k_y k_z & \left(-k_x^2 - k_y^2 + \frac{\omega^2 n_z^2}{c^2}\right) \end{vmatrix} = 0 \qquad (6)$$

Evaluating the determinant of eqn (6), and rearranging the terms, we obtain

$$\frac{\omega^4}{c^4} - \frac{\omega^2}{c^2}\left(\frac{k_x^2 + k_y^2}{n_z^2} + \frac{k_x^2 + k_z^2}{n_y^2} + \frac{k_y^2 + k_z^2}{n_x^2}\right) +$$

$$\left(\frac{k_x^2}{n_y^2 n_z^2} + \frac{k_y^2}{n_x^2 n_z^2} + \frac{k_z^2}{n_x^2 n_y^2}\right)(k_x^2 + k_y^2 + k_z^2) = 0$$

In the case of a uniaxial material, choosing the optic axis to be in the z direction so that $n_x = n_y = n_o$ and $n_z = n_e$, this expression can be factored into

$$\left(\frac{k_x^2}{n_o^2} + \frac{k_y^2}{n_o^2} + \frac{k_z^2}{n_o^2} - \frac{\omega^2}{c^2}\right)\left(\frac{k_x^2}{n_e^2} + \frac{k_y^2}{n_e^2} + \frac{k_z^2}{n_o^2} - \frac{\omega^2}{c^2}\right) = 0. \qquad (8)$$

Setting either of the factors in eqn 8 to zero will define an ellipsoidal surface in space of allowed wave vectors k. The first factor being zero defines a sphere corresponding to ordinary rays, in which the effective refractive index is exactly $n_o$. The second defines a spheroid symmetric about the z axis. This solution corresponds to extraordinary rays in which the effective refractive index is in between $n_o$ and $n_e$. Therefore for any arbitrary direction of propagation, two distinct wavevectors k are allowed corresponding to the polarizations of the ordinary and extraordinary rays. A general state of polarization launched into the medium can be decomposed into two such waves which will then propagate with different k vectors (except in the case of propagation in the direction of the optic axis). For a biaxial material a similar but somewhat more complicated condition on the two waves can be described.

# 3

# Nonlinear Optics

Nonlinear optics (NLO) is the branch of optics that describes the behaviour of light in *nonlinear media*, that is, media in which the dielectric polarization P responds nonlinearly to the electric field E of the light. This nonlinearity is typically only observed at very high light intensities (values of the electric field comparable to interatomic electric fields, typically $10^8$ V/m) such as those provided by pulsed lasers. Above the Schwinger limit, the vacuum itself is expected to become nonlinear. In nonlinear optics, the superposition principle no longer holds.

Nonlinear optics remained unexplored until the discovery of Second harmonic generation shortly after demonstration of the first laser. (Peter Franken *et al.* at University of Michigan in 1961)

## Nonlinear Optical Processes

Nonlinear optics gives rise to a host of optical phenomena:

### *Frequency Mixing Processes*

- Second harmonic generation (SHG), or *frequency doubling*, generation of light with a doubled frequency (half the wavelength), two photons are destroyed creating a single photon at two times the frequency.
- Third harmonic generation (THG), generation of light with a tripled frequency (one-third the wavelength), three photons are destroyed creating a single photon at three times the frequency.
- High harmonic generation (HHG), generation of light with frequencies much greater than the original (typically 100 to 1000 times greater)

- Sum frequency generation (SFG), generation of light with a frequency that is the sum of two other frequencies (SHG is a special case of this)
- Difference frequency generation (DFG), generation of light with a frequency that is the difference between two other frequencies
- Optical parametric amplification (OPA), amplification of a signal input in the presence of a higher-frequency pump wave, at the same time generating an *idler* wave (can be considered as DFG)
- Optical parametric oscillation (OPO), generation of a signal and idler wave using a parametric amplifier in a resonator (with no signal input)
- Optical parametric generation (OPG), like parametric oscillation but without a resonator, using a very high gain instead
- Spontaneous parametric down conversion (SPDC), the amplification of the vacuum fluctuations in the low gain regime
- Optical rectification (OR), generation of quasi-static electric fields.
- Nonlinear light-matter interaction with free electrons and plasmas

### *Other Nonlinear Processes*

- Optical Kerr effect, intensity dependent refractive index (a $\chi^{(3)}$ effect)
  - o Self-focusing, an effect due to the Optical Kerr effect (and possibly higher order nonlinearities) caused by the spatial variation in the intensity creating a spatial variation in the refractive index
  - o Kerr-lens modelocking (KLM), the use of Self-focusing as a mechanism to mode lock laser.
  - o Self-phase modulation (SPM), an effect due to the Optical Kerr effect (and possibly higher order nonlinearities) caused by the temporal variation in the intensity creating a temporal variation in the refractive index
  - o Optical solitons, An equilibrium solution for either an optical pulse (temporal soliton) or Spatial mode (spatial soliton) that does not change during propagation due to a balance between diffraction and the Kerr effect (e.g. Self-phase

modulation for temporal and Self-focusing for spatial solitons).

- Cross-phase modulation (XPM)
- Four-wave mixing (FWM), can also arise from other nonlinearities
- Cross-polarized wave generation (XPW), a $\chi^{(3)}$ effect in which a wave with polarization vector perpendicular to the input one is generated
- Modulational instability
- Raman amplification
- Optical phase conjugation
- Stimulated Brillouin scattering, interaction of photons with acoustic phonons
- Multi-photon absorption, simultaneous absorption of two or more photons, transferring the energy to a single electron
- Multiple photoionisation, near-simultaneous removal of many bound electrons by one photon

### *Related Processes*

In these processes, the medium has a linear response to the light, but the properties of the medium are affected by other causes:

- Pockels effect, the refractive index is affected by a static electric field; used in electro-optic modulators;
- Acousto-optics, the refractive index is affected by acoustic waves (ultrasound); used in acousto-optic modulators.
- Raman scattering, interaction of photons with optical phonons;

## Parametric Processes

Nonlinear effects fall into two qualitatively different categories, parametric and non-parametric effects. A parametric non-linearity is an interaction in which the quantum state of the nonlinear material is not changed by the interaction with the optical field. As a consequence of this, the process is 'instantaneous'; Energy and momentum conserving in the optical field, making phase matching important; and polarization dependent.

### *Theory*

Parametric and lossy 'instantaneous' (i.e. electronic) nonlinear optical phenomena, in which the optical fields are not too large, can be described by a Taylor series expansion of the dielectric Polarization density (dipole moment per unit volume) P($t$) at time $t$ in terms of the electrical field E($t$):

$$P(t) = \varepsilon_0(\chi^{(1)}E(t) + \chi^{(2)}E^2(t) + \chi^{(3)}E^3(t) + \ldots).$$

Here, the coefficients $\chi^{(n)}$ are the $n$-th order susceptibilities of the medium and the presence of such a term is generally referred to as an $n$-th order nonlinearity. In general $\chi^n$ is an $n+1$ order tensor representing both the polarization dependent nature of the parametric interaction as well as the symmetries (or lack thereof) of the nonlinear material.

### *Wave-equation in a Nonlinear Material*

Central to the study of electromagnetic waves is the wave equation. Starting with Maxwell's equations in an isotropic space containing no free charge, it can be shown that:

$$\nabla \times \nabla \times E + \frac{n^2}{c^2}\frac{\partial^2}{\partial t^2}E = -\frac{1}{\varepsilon_0 c^2}\frac{\partial^2}{\partial t^2}P^{NL},$$

where $P^{NL}$ is the nonlinear part of the Polarization density and n is the refractive index which comes from the linear term in P.

Note one can normally use the vector identity

$$\nabla \times (\nabla \times V) = \nabla(\nabla \cdot V) - \nabla^2 V$$

and Gauss's law,

$$\nabla \cdot D = 0,$$

to obtain the more familiar wave equation

$$\nabla^2 E - \frac{n^2}{c^2}\frac{\partial^2}{\partial t^2}E = 0.$$

For nonlinear medium Gauss's law does not imply that the identity

$$\nabla \cdot E = 0$$

is true in general, even for an isotropic medium. However even when this term is not identically 0, it is often negligibly small and thus in practice is usually ignored giving us the standard nonlinear wave-equation:

$$\nabla^2 E - \frac{n^2}{c^2}\frac{\partial^2}{\partial t^2}E = \frac{1}{\varepsilon_0 c^2}\frac{\partial^2}{\partial t^2}P^{NL}.$$

### *Nonlinearities as a Wave Mixing Process*

The nonlinear wave-equation is an inhomogeneous differential equation. The general solution comes from the study of ordinary differential equations and can be solved by the use of a Green's function. Physically one gets the normal electromagnetic wave solutions to the homogeneous part of the wave equation:

$$\nabla^2 \mathrm{E} - \frac{n^2}{c^2}\frac{\partial^2}{\partial t^2}\mathrm{E} = 0.$$

and the inhomogenous term

$$\frac{1}{\varepsilon_0 c^2}\frac{\partial^2}{\partial t^2}\mathrm{P}^{NL},$$

acts as a driver/source of the electromagnetic waves. One of the consequences of this is a nonlinear interaction that will result in energy being mixed or coupled between different frequencies which is often called a 'wave mixing'.

In general an $n$-th order will lead to $n+1$-th wave mixing. As an example, if we consider only a second order nonlinearity (three-wave mixing), then the polarization, $P$, takes the form

$$\mathrm{P}^{NL} = \varepsilon_0 \chi^{(2)} \mathrm{E}^2(t).$$

If we assume that $E(t)$ is made up of two components at frequencies $\omega_1$ and $\omega_2$, we can write $E(t)$ as

$$\mathrm{E}(t) = E_1 e^{-i\omega_1 t} + E_2 e^{-i\omega_2 t} + c.c.$$

where *c.c.* stands for complex conjugate. Plugging this into the expression for $P$ gives

$$\begin{aligned}\mathrm{P}^{NL} = \varepsilon_0 \chi^{(2)} \mathrm{E}^2(t) \quad &= \varepsilon_0 \chi^{(2)} [|E_1|^2 e^{-i2\omega_1 t} + |E_2|^2 e^{-i2\omega_2 t} \\ &+ 2E_1 E_2 e^{-i(\omega_1+\omega_2)t} \\ &+ 2E_1 E_2^* e^{-i(\omega_1-\omega_2)t} \\ &+ 2(|E_1| + |E_2|)e^0],\end{aligned}$$

which has frequency components at $2\omega_1, 2\omega_2$, $\omega_1+\omega_2$, $\omega_1-\omega_2$, and 0. These three-wave mixing processes correspond to the nonlinear effects known as second harmonic generation, sum frequency generation, difference frequency generation and optical rectification respectively.

***Note:*** Parametric generation and amplification is a variation of difference frequency generation, where the lower-frequency of one of the two generating fields is much weaker (parametric amplification) or completely absent (parametric generation). In the latter case, the fundamental quantum-mechanical uncertainty in the electric field initiates the process.

### Phase Matching

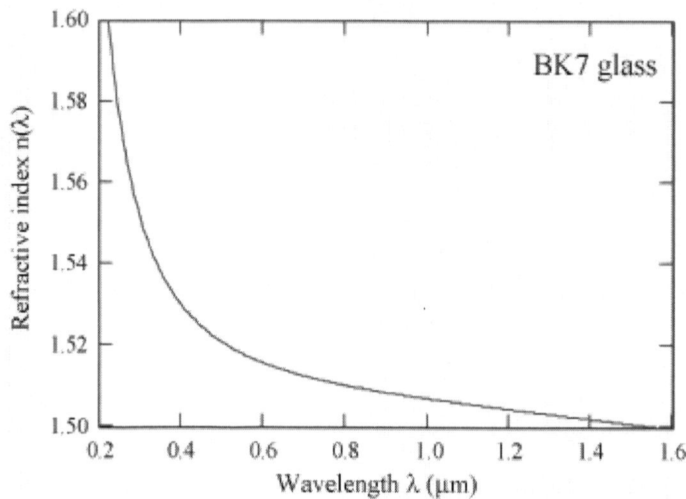


***Figure:** Most transparent materials, like the BK7 glass shown here, have normal dispersion: The index of refraction decreases monotonically as a function of wavelength (or increases as a function of frequency). This makes phase-matching impossible in most frequency-mixing processes. For example, in SHG, there is no simultaneous solution to $\omega' = 2\omega$ and k'=2k in these materials. Birefringent materials avoid this problem by having two indices of refraction at once.*

The above ignores the position dependence of the electrical fields. In a typical situation, the electrical fields are travelling waves described by

$$E_j(\mathrm{x},t) = e^{i(\mathrm{k}_j \cdot \mathrm{x} - \omega_j t)} + c.c.,$$

at position $\mathrm{x}$, with the wave vector $\|\mathrm{k}_j\| = n(\omega_j)\omega_j / c$, where $c$ is the velocity of light and $n(\omega_j)$ the index of refraction of the medium at angular frequency $\omega_j$. Thus, the second-order polarization at angular frequency $\omega_3 = \omega_1 + \omega_2$ is

$$P^{(2)}(\mathrm{x},t) \propto E_1^{n_1} E_2^{n_2} e^{i((\mathrm{k}_1+\mathrm{k}_2)\cdot \mathrm{x} - \omega_3 t)} + c.c.$$

At each position $\mathrm{x}$ within the nonlinear medium, the oscillating second-order polarization radiates at angular frequency $\omega_3$ and a corresponding wave vector $\|\mathrm{k}_j\| = n(\omega_j)\omega_j / c$. Constructive interference, and therefore a high intensity $\omega_3$ field, will occur only if

$$\mathrm{k}_3 = \mathrm{k}_1 + \mathrm{k}_2.$$

The above equation is known as the *phase matching condition.* Typically, three-wave mixing is done in a birefringent crystalline material (I.e., the refractive index depends on the polarization and direction of the light that passes through.), where the polarizations of the fields and the orientation of the crystal are chosen such that the phase-matching condition is fulfilled. This phase matching technique is called angle tuning. Typically a crystal has three axes, one or two of which have a different refractive index than the other one(s). Uniaxial crystals, for example, have a single preferred axis, called the extraordinary (e) axis, while the other two are ordinary axes (o). There are several schemes of choosing the polarizations for this crystal type. If the signal and idler have the same polarization, it is called "Type-I phase-matching", and if their polarizations are perpendicular, it is called "Type-II phase-matching". However, other conventions exist that specify further which frequency has what polarization relative to the crystal axis. These types are listed below, with the convention that the signal wavelength is shorter than the idler wavelength.

***Table:*** *Phase-matching types*

| *Polarizations* | | | *Scheme* |
|---|---|---|---|
| ***Pump*** | ***Signal*** | ***Idler*** | |
| e | o | o | Type I |
| e | o | e | Type II (or IIA) |
| e | e | o | Type III (or IIB) |
| e | e | e | Type IV |
| o | o | o | Type V (or Type-0 or Zero) |
| o | o | e | Type VI (or IIB or IIIA) |
| o | e | o | Type VII (or IIA or IIIB) |
| o | e | e | Type VIII (or I) |

Most common nonlinear crystals are negative uniaxial, which means that the *e* axis has a smaller refractive index than the *o* axes. In those crystals, type I and II phasematching are usually the most suitable schemes. In positive uniaxial crystals, types VII and VIII are more suitable. Types II and III are essentially equivalent, except that the names of signal and idler are swapped when the signal has a longer wavelength than the idler. For this reason, they are sometimes called IIA and IIB. The type numbers V–VIII are less common than I and II and variants.

One undesirable effect of angle tuning is that the optical frequencies involved do not propagate collinearly with each other. This is due to the fact that the extraordinary wave propagating through a birefringent crystal possesses a Poynting vector that is not parallel with the propagation vector. This would lead to beam walk-off which limits the nonlinear optical conversion efficiency. Two other methods of phase matching avoids beam walk-off by forcing all frequencies to propagate at a 90 degree angle with respect to the optical axis of the crystal. These methods are called temperature tuning and quasi-phase-matching.

Temperature tuning is where the pump (laser) frequency polarization is orthogonal to the signal and idler frequency polarization. The birefringence in some crystals, in particular Lithium Niobate is highly temperature dependent. The crystal is controlled at a certain temperature to achieve phase matching conditions.

The other method is quasi-phase matching. In this method the frequencies involved are not constantly locked in phase with each other, instead the crystal axis is flipped at a regular interval Ë, typically 15 micrometres in length. Hence, these crystals are called periodically poled. This results in the polarization response of the crystal to be shifted back in phase with the pump beam by reversing the nonlinear susceptibility. This allows net positive energy flow from the pump into the signal and idler frequencies. In this case, the crystal itself provides the additional wavevector $k=2\pi/\lambda$ (and hence momentum) to satisfy the phase matching condition. Quasi-phase matching can be expanded to chirped gratings to get more bandwidth and to shape an SHG pulse like it is done in a dazzler. SHG of a pump and Self-phase modulation (emulated by second order processes) of the signal and an optical parametric amplifier can be integrated monolithically.

## Higher-order Frequency Mixing

The above holds for $\chi^{(2)}$ processes. It can be extended for processes where $\chi^{(3)}$ is nonzero, something that is generally true in any medium without any symmetry restrictions. Third-harmonic generation is a $\chi^{(3)}$ process, although in laser applications, it is usually implemented as a two-stage process: first the fundamental laser frequency is doubled and then the doubled and the fundamental frequencies are added in a sum-frequency process. The Kerr effect can be described as a $\chi^{(3)}$ as well.

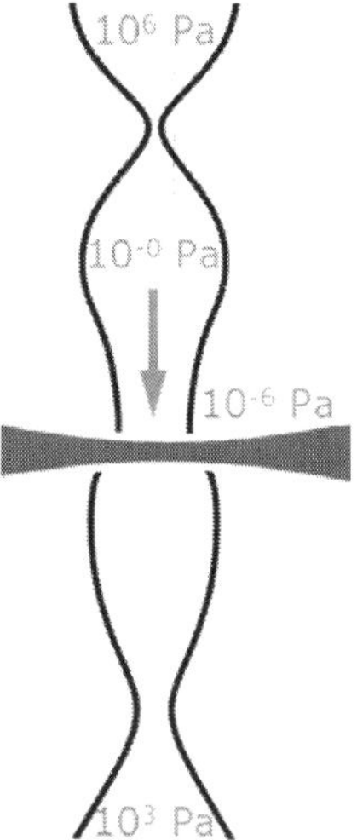


At high intensities the Taylor series, which led the domination of the lower orders, does not converge anymore and instead a time based model is used. When a noble gas atom is hit by an intense laser pulse, which has an electric field strength comparable to the Coulomb field of the atom, the outermost electron may be ionized from the atom. Once freed, the electron can be accelerated by the electric field of the light, first moving away from the ion, then back towards it as the field changes direction. The electron may then recombine with the ion, releasing its energy in the form of a photon. The light is emitted at every peak of the laser light field which is intense enough, producing a series of attosecond light flashes. The photon energies generated by this process can extend past the 800th harmonic order up to 1300 eV. This is called high-order harmonic generation. The laser must be linearly polarized, so that the electron returns to the vicinity of the parent ion. High-order harmonic generation has been observed in noble gas jets, cells, and gas-filled capillary waveguides.

### Example uses of Nonlinear Optics

***Frequency Doubling:*** One of the most commonly used frequency-mixing processes is frequency doubling or second-harmonic generation. With this technique, the 1064-nm output from Nd:YAG lasers or the 800-nm output from Ti:sapphire lasers can be converted to visible light, with wavelengths of 532 nm (green) or 400 nm (violet), respectively.

Practically, frequency-doubling is carried out by placing a nonlinear medium in a laser beam. While there are many types of nonlinear media, the most common media are crystals. Commonly used crystals are BBO (β-barium borate), KDP (potassium dihydrogen phosphate), KTP (potassium titanyl phosphate), and lithium niobate. These crystals have the necessary properties of being strongly birefringent, having a specific crystal symmetry and of course being transparent for both the impinging laser light and the frequency doubled wavelength, and have high damage thresholds which make them resistant against the high-intensity laser light. However, organic polymeric materials are set to take over from crystals as they are cheaper to make, have lower drive voltages and superior performance.

## Optical Phase Conjugation

It is possible, using nonlinear optical processes, to exactly reverse the propagation direction and phase variation of a beam of light. The reversed beam is called a *conjugate* beam, and thus the technique is known as optical phase conjugation (also called *time reversal, wavefront reversal* and *retroreflection*).

One can interpret this nonlinear optical interaction as being analogous to a real-time holographic process. In this case, the interacting beams simultaneously interact in a nonlinear optical material to form a dynamic hologram (two of the three input beams), or real-time diffraction pattern, in the material. The third incident beam diffracts off this dynamic hologram, and, in the process, reads out the *phase-conjugate* wave. In effect, all three incident beams interact (essentially) simultaneously to form several real-time holograms, resulting in a set of diffracted output waves that phase up as the "time-reversed" beam. In the language of nonlinear optics, the interacting beams result in a nonlinear polarization within the material, which coherently radiates to form the phase-conjugate wave.

The most common way of producing optical phase conjugation is to use a four-wave mixing technique, though it is also possible to use processes such as stimulated Brillouin scattering. A device producing the phase conjugation effect is known as a phase conjugate mirror (PCM).

For the four-wave mixing technique, we can describe four beams ($j$ = 1,2,3,4) with electric fields:

$$\Xi_j(\mathrm{x},t) = \frac{1}{2}E_j(\mathrm{x})e^{i(\omega_j t - \mathrm{k}\cdot\mathrm{x})} + \text{c.c.}$$

where $E_j$ are the electric field amplitudes. $\Xi_1$ and $\Xi_2$ are known as the two pump waves, with $\Xi_3$ being the signal wave, and $\Xi_4$ being the generated conjugate wave.

If the pump waves and the signal wave are superimposed in a medium with a non-zero $\chi^{(3)}$, this produces a nonlinear polarization field:

$$P_{NL} = \epsilon_0 \chi^{(3)} (\Xi_1 + \Xi_2 + \Xi_3)^3$$

resulting in generation of waves with frequencies given by $\omega = \pm\omega_1 \pm\omega_2 \pm\omega_3$ in addition to third harmonic generation waves with $\omega = 3\omega_1$, $3\omega_2$, $3\omega_3$.

As above, the phase-matching condition determines which of these waves is the dominant. By choosing conditions such that $\omega = \omega_1 + \omega_2 - \omega_3$ and $k = k_1 + k_2 - k_3$, this gives a polarization field:

$$P_\omega = \frac{1}{2}\chi^{(3)}\epsilon_0 E_1 E_2 E_3^* e^{i(\omega t - k\cdot x)} + \text{c.c.}.$$

This is the generating field for the phase conjugate beam, $\Xi_4$. Its direction is given by $k_4 = k_1 + k_2 - k_3$, and so if the two pump beams are counterpropagating ($k_1 = -k_2$), then the conjugate and signal beams propagate in opposite directions ($k_4 = -k_3$). This results in the retroreflecting property of the effect.

Further, it can be shown for a medium with refractive index $n$ and a beam interaction length $l$, the electric field amplitude of the conjugate beam is approximated by

$$E_4 = \frac{i\omega l}{2nc}\chi^{(3)} E_1 E_2 E_3^*$$

(where $c$ is the speed of light). If the pump beams $E_1$ and $E_2$ are plane (counterpropagating) waves, then:

$$E_4(x) \propto E_3^*(x);$$

that is, the generated beam amplitude is the complex conjugate of the signal beam amplitude. Since the imaginary part of the amplitude contains the phase of the beam, this results in the reversal of phase property of the effect.

Note that the constant of proportionality between the signal and conjugate beams can be greater than 1. This is effectively a mirror with a reflection coefficient greater than 100%, producing an amplified reflection. The power for this comes from the two pump beams, which are depleted by the process.

The frequency of the conjugate wave can be different from that of the signal wave. If the pump waves are of frequency $\omega_1 = \omega_2 = \omega$, and the signal wave higher in frequency such that $\omega_3 = \omega + \Delta\omega$, then the conjugate wave is of frequency $\omega_4 = \omega - \Delta\omega$. This is known as *frequency flipping.*

## Refractometry

Refractometry is the method of measuring substances' refractive index (one of their fundamental physical properties) in order to, for example, assess their composition or purity. A refractometre is the instrument used to measure refractive index ("RI"). Although refractometres are best known for measuring liquids, they are also used to measure gases and solids; such as glass and gemstones.

The RI of a substance is strongly influenced by temperature and the wavelength of light used to measure it, therefore, care must be taken to control or compensate for temperature differences and wavelength. RI measurements are usually reported at a reference temperature of 20 degrees Celsius, which is equal to 68 degrees Fahrenheit, and considered to be room temperature. A reference wavelength of 589.3 nm (the sodium D line) is most often used. Though RI is a dimensionless quantity, it is typically reported as nD20, where the "n" represents refractive index, the "D" denotes the wavelength, and the 20 denotes the reference temperature. Therefore, the refractive index of water at 20 degrees Celsius, taken at the Sodium D Line, would be reported as 1.3330 nD20.

Refractometres are frequently used by grape growers and kiwifruit growers for Brix testing of sucrose levels in their fruit. Refractometry is also used in the gelatin industry. To convert the RI of a gelatin sol (reported in Brix) to a gelatin concentration, one need only multiply by eight-tenths (0.8). A sol with a 10.0 RI would therefore be 8% gelatin by weight. This is known to be a reliable conversion for gelatin sols as low as 1% up to over 50%.

### *Refractometre*

A refractometre is a laboratory or field device for the measurement of an index of refraction (refractometry). The index of refraction is calculated from Snell's law and can be calculated from the composition of the material using the Gladstone–Dale relation.

## Types of Refractometres

There are four main types of refractometres: traditional handheld refractometres, digital handheld refractometres, laboratory or Abbe refractometres, and inline process refractometres. There is also the Rayleigh Refractometre used (typically) for measuring the refractive indices of gases.

In veterinary medicine, a refractometre is used to measure the total plasma protein in a blood sample and urine specific gravity.

In drug diagnostics, a refractometre is used to measure the specific gravity in human urine.

In gemology, a refractometre is used to help identify gem materials by measuring their refractive index. Gemstones are transparent minerals and can therefore be examined using optical methods. As the refractive index is a material constant dependent on the chemical composition of a substance, it provides information on the type and quality of a gemstone. Classification with a special gemstone refractometre is an easy-to-use method with which the authenticity and quality of a stone can be evaluated. The gemstone refractometre is therefore a piece of basic equipment in a gemological laboratory. Due to the dependence of the refractive index (dispersion) on the wavelength of the light used, the measurement is normally taken at the wavelength of the sodium-D-line (NaD) of 589 nm. This is either filtered out from daylight or generated with a monochromatic light-emitting diode (LED). Certain stones such as rubies, sapphires, tourmalines and topaz are optically anisotropic. They demonstrate birefringence based on the polarisation plane of the light. The two different refractive indexes are classified using a polarisation filter. Gemstone refractometres are available both as classic optical instruments and as electronic measurement devices with a digital display.

In marine aquarium keeping, a refractometre is used to measure the salinity and specific gravity of the water.

In homebrewing, a refractometre is used to measure the specific gravity before fermentation to determine the amount of fermentable sugars which will potentially be converted to alcohol.

## Automatic Refractometres

Automatic refractometres automatically measure the refractive index of a sample. The automatic measurement of the refractive index of the sample is based on the determination of the critical angle of total reflection. A light source, usually a long-life LED, is focused onto a prism surface via a lens system. An interference filter guarantees the specified wavelength. Due to focusing light to a spot at the prism surface, a wide range of different angles is covered. As shown in the figure *"Schematic setup of an automatic refractometre"* the measured sample is in direct contact with the measuring prism. Depending on its refractive index, the incoming light below the critical angle of total reflection is partly transmitted into the sample, whereas for higher angles of incidence the light is totally reflected. This dependence of the reflected light intensity from the incident angle is measured with a high-resolution sensor array. From the video signal taken with the CCD sensor the refractive index of the sample can be calculated. This method of detecting the angle of total reflection is independent on the sample properties. It is even possible to measure the refractive index of optical dens strongly absorbing samples or samples containing air bubbles or solid particles . Furthermore, only a few microliters are required and the sample can be recovered. This determination of the refraction angle is independent of vibrations and other environmental disturbances.

### *Influence of Wavelength*

The refractive index of a certain sample varies for nearly all materials for different wavelengths. This dispersion relation is characteristic for every material. In the visible wavelength range a decrease of the refractive index and nearly no absorption is observable. In the infrared wavelength range several absorption maxima and fluctuations in the refractive index appear. To guarantee a high quality measurement with an accuracy of up to 0.00002 in the refractive index the wavelength has to be determined correctly. Therefore, in modern refractometres the wavelength is tuned to a bandwidth of +/-0.2 nm to ensure correct results for samples with different dispersions.

### *Influence of Temperature*

Temperature has a very important influence on the refractive index measurement . Therefore, the temperature of the prism and the temperature of the sample have to be controlled with high precision. There are several subtly different designs for controlling the temperature but there are some key factors common to all such as high precision temperature sensors and Peltier devices to control the temperature of the sample and the prism. The temperature control accuracy of these devices should be designed so that the variation in sample temperature is small enough that it will not cause a detectable refractive index change. External water baths were used in the past but are no longer needed.

### *Extended Possibilities of Automatic Refractometres*

Automatic refractometres are microprocessor-controlled electronic devices. This means they can have a high degree of automation and also be combined with other measuring devices.

### Flow Cells

There are different types of sample cells available, ranging from a flow cell for a few microliters to sample cells with a filling funnel for fast sample exchange without cleaning the measuring prism in between. The sample cells can also be used for the measurement of poisonous and toxic samples with minimum exposure to the sample. Micro cells require only a few microliters volume, assure good recovery of expensive samples and prevent evaporation of volatile samples or solvents. They can also be used in automated systems for automatic filling of the sample onto the refractometre prism. For convenient filling of the sample through a funnel, flow cells with a filling funnel are available. These are used for fast sample exchange in quality control applications.

### Automatic Sample Feeding

Once an automatic refractometre is equipped with a flow cell, the sample can either be filled by means of a syringe or by using a peristaltic pump. Modern refractometres have the option of a built-in peristaltic pump. This is controlled via the instrument‘s software menu. A peristaltic pump opens the way to monitor batch processes in the laboratory or perform multiple measurements on one sample without any user interaction. This eliminates human error and assures a high sample throughput.

If an automated measurement of a large number of samples is required, modern automatic refractometres can be combined with an automatic sample changer. The sample changer is controlled by the refractometre and assures fully automated measurements of the samples placed in the vials of the sample changer for measurements.

### Multiparametre Measurements

Today's laboratories do not only want to measure the refractive index of samples, but several additional parametres like density or viscosity to perform efficient quality control. Due to the microprocessor control and a number of interfaces, automatic refractometres are able to communicate with computers or other measuring devices, e.g. density metres, pH metres or viscosity metres, to store refractive index data and density data (and other parametres) into one database.

### Software Features

Automatic refractometres do not only measure the refractive index, but offer a lot of additional software features, like:

- Instrument settings and configuration via software menu
- Automatic data recording into a database
- User-configurable data output
- Export of measuring data into Mirosoft Excel data sheets
- Statistical functions
- Predefined methods for different kinds of applications
- Automatic checks and adjustments
- Check if sufficient amount of sample is on the prism
- Data recording only if the results are plausible

### *Pharma Documentation and Validation*

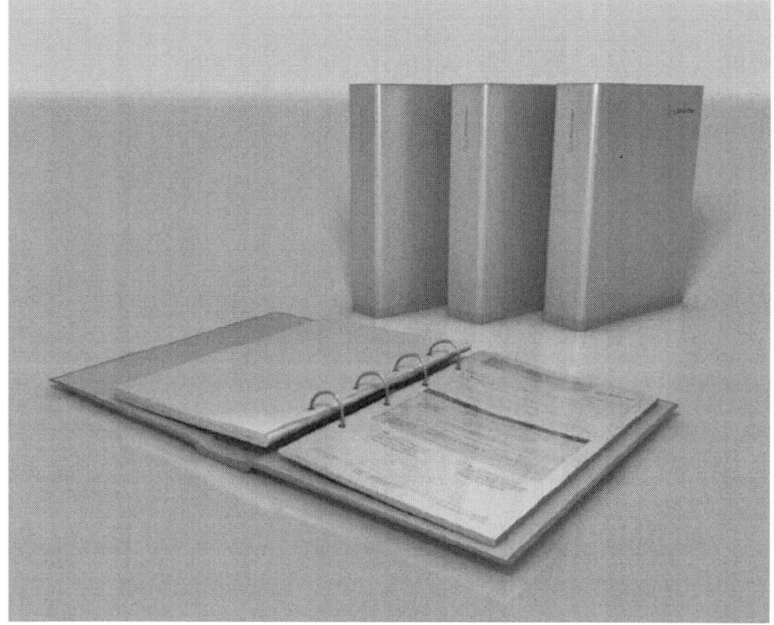

***Figure:*** *Typical Pharma Validation and Qualification Folder*

Refractometres are often used in pharmaceutical applications for quality control of raw intermediate and final products. The manufacturers of pharmaceuticals have to follow several international regulations like FDA 21 CFR Part 11, GMP, Gamp 5, USP<1058>, which require a lot of documentation work. The manufacturers of automatic refractometres support these users providing instrument software fulfills the requirements of 21 CFR Part 11, with user levels, electronic signature and audit trail. Furthermore, Pharma Validation and Qualification Packages are available containing:

- Qualification Plan (QP)
- Design Qualification (DQ)
- Risk Analysis
- Installation Qualification (IQ)
- Operational Qualification (OQ)
- Check List 21 CFR Part 11 / SOP
- Performance Qualification (PQ)

## Auguste Michel-Lévy

Auguste Michel-Lévy (7 August 1844 - 27 September 1911) was a French geologist. He was born in Paris.

He became inspector-general of mining | mines, and director of the Geological Survey of France. He was distinguished for his researches on extrusive rocks, their microscopic structure and origin; and he employed the polarizing microscope early on for the determination of minerals. In his many contributions to scientific journals he described the granulite group, and dealt with pegmatites, variolites, eurites, the ophites of the Pyrenees, the extinct volcanoes of Central France, gneisses, and the origin of crystalline schists.

He wrote *Structures et classification des roches éruptives* (1889), but his more elaborate studies were carried on with F Fouqué. Together they wrote on the artificial production of feldspar, nepheline and other minerals, and also of meteorites, and produced *Minéralogie micrographique* (1879) and *Synthése des minéraux et des roches* (1882). Levy also collaborated with Alfred Lacroix in *Les Minéraux des roches* (1888) and *Tableau des minéraux des roches* (1889).

Michel-Lévy pioneered the use of birefringence to identify minerals in thin section with a petrographic microscope. He is widely known for the Michel-Lévy interference colour chart, which defines the interference colours from different orders of birefringence.

He also created classification schemes for igneous rocks which accounted their mineralogy, texture, and composition, and showed that igneous rocks of different mineralogies could be formed from the same chemical composition, with different conditions of crystallization.

## Interference Colour Chart

An interference colour chart, first developed by Auguste Michel-Lévy, is an optical mineralogy tool to identify minerals in thin section using a petrographic microscope. With a known thickness of the thin section, minerals have specific and predictable colours in cross-polarized light, and this chart can help identify minerals. The colours are produced by the difference in speed in the fast and slow rays, also known as birefringence.

When using the chart, it is important to remember these tips:

- Isotropic and opaque (metallic) minerals cannot be identified this way.
- Rotate the stage in order to get the maximum colour, and therefore, the maximum birefringence.
- Each mineral, depending on the orientation, may not exhibit the maximum birefringence. It is important to try and find multiple like minerals in order to get the best value of birefringence.
- Uniaxial minerals can look isotropic (always extinct) if the mineral is cut perpendicular to the optic axis. (But this situation can be revealed with the Conoscopic interference pattern).

## Pleochroism

Pleochroism is an optical phenomenon in which a substance appears to be different colours when observed at different angles, especially with polarized light.

### *Background*

Anisotropic crystals will have optical properties that vary with the direction of light. The polarization of light determines the direction of the electric field, and crystals will respond in different ways if this angle is changed. These kinds of crystals have one or two optical axes. If absorption of light varies with the angle relative to the optical axis in a crystal then pleochroism results.

Anisotropic crystals have double refraction of light where light of different polarizations is bent different amounts by the crystal, and therefore follows different paths through the crystal. The components of a divided light beam follow different paths within the mineral and travel at different speeds. When the mineral is observed at some angle, light following some combination of paths and polarizations will be present, each of which will have had light of different colours absorbed. At another angle, the light passing through the crystal will be composed of another combination of light paths and polarizations, each with their own colour. The light passing through the mineral will therefore have different colours when it is viewed from different angles, making the stone seem to be of different colours.

Tetragonal, trigonal and hexagonal minerals can only show two colours and are called dichroic. Orthorhombic, monoclinic and triclinic crystals can show three and are trichroic. For example hypersthene, with two optical axes, can have red, yellow or blue appearance when oriented in three different ways in three dimensional space. Isometric minerals cannot exhibit pleochroism. Tourmaline is notable for exhibiting strong pleochroism. Gems are sometimes cut and set either to display pleochroism or to hide it, depending on the colours and their attractiveness.

The pleochroic colours are at their maximum when light is polarized parallel with a crystallographic axis. The axes are designated X, Y and Z. These axes can be determined from the appearance of a crystal in a conoscopic interference pattern. Where there are two optical axes, the acute bisection of the axes gives Z for positive minerals and X for negative minerals and the obtuse bisection give the alternative axis (X or Z). Perpendicular to these is the Y axis. The colour is measured with the polarization parallel to each direction. An absorption formula records the amount of absorption parallel to each axis in the form of $X < Y < Z$ with the left most having the least absorption and the rightmost the most.

### *In Mineralogy*

Pleochroism is an extremely useful tool in mineralogy for mineral identification, since minerals that are otherwise very similar often have very different pleochroic colour schemes. In such cases, a thin section of the mineral is used and examined under polarized transmitted light with a petrographic microscope.

### *List of Pleochroic Minerals*

Purple and violet:

- Amethyst (very low): purple / purple
- Andalusite (strong): green brown / dark red / purple
- Beryl (medium): purple / colourless
- Corundum (high): purple / orange
- Hypersthene (strong): purple/orange
- Spodumene (Kunzite) (strong): purple / purple / clear / pink
- Tourmaline (strong): pale purple / purple

### *Blue*

- Aquamarine (medium): colourless-light blue / light blue, dark blue

- Alexandrite (strong): Dark red-purple/orange/green
- Apatite (strong): blue-yellow/blue-colourless
- Benitoite (strong): colourless / dark blue
- Cordierite (very strong): orthorhombic blue brown / yellow / greenish brown / gray blue / blue to purple
- Corundum (strong): violet-dark blue / light blue-green
- Iolite (strong): colourless / yellow / blue / dark blue-violet
- Topaz (very low): colourless / pale blue / pink
- Tourmaline (strong): dark blue / light blue
- Zoisite (strong): blue / red purple / yellow green
- Zircon (strong): blue / clear / gray

### *Green*

- Alexandrite (strong): dark red / orange / green
- Andalusite (strong): brown green / dark red
- Corundum (strong): green / yellow green
- Emerald (strong): Green / Blue Green
- Peridot (low): yellow-green / green / colourless
- Sphene (medium): brown green / blue green
- Tourmaline (strong): blue green / brown green / yellow green
- Zircon (low): greenish brown / green

### *Yellow*

- Citrine (very weak): pale yellow / pale yellow
- Chrysoberyl (very weak): red-yellow/yellow-green/green
- Corundum (weak): yellow / pale yellow
- Danburite (weak): very pale yellow / pale yellow
- Orthoclase (weak): pale yellow / pale yellow
- Phenacite (medium): colourless / yellow orange
- Spodumene (medium): pale yellow / pale yellow
- Topaz (medium): tan / yellow / yellow orange
- Tourmaline (medium): pale yellow / dark yellow
- Hornblende (strong) light green/dark green/yellow/brown

### *Brown and Orange*

- Corundum (strong): yellow brown / orange
- Topaz (medium): brown-yellow/brown yellow dull

- Tourmaline (very low): dark brown / light brown
- Zircon (very weak): brown red/brown-yellow
- Biotite (medium): brown

### *Red and Pink*

- Alexandrite (strong): dark red / orange / green
- Andalusite (strong): dark red / brown red
- Corundum (strong): violet red / orange red
- Morganite (medium): light red / red violet
- Tourmaline (strong): dark red / light red
- Zircon (medium): purple / red brown

### *Extinction (Optical Mineralogy)*

Extinction is a term used in optical mineralogy and petrology, which describes when cross-polarized light dims, as viewed through a thin section of a mineral in a petrographic microscope. Isotropic minerals, opaque (metallic) minerals, or amophous materials (glass) show no light (i.e. constant extinction). Anisotropic minerals will show one extinction for each 90 degrees of stage rotation.

The extinction angle is the measure between the cleavage direction or habit of a mineral and the extinction. To find this, simply line up the cleavage lines/long direction with one of the cross hairs in the microscope, and turn the mineral until the extinction occurs. The number of degrees the stage was rotated is the extinction angle, between 0-89 degrees. 90 degrees would be considered zero degrees, and is known as parallel extinction. Inclined extinction is a measured angle between 1-89 degrees. Minerals with two cleavages can have two extinction angles, and minerals in which the multiple angles are the same are called symmetrical extinction. Minerals that have no cleavage or elongation can not have an extinction angle.

Minerals with undulose extinction, solid solution/zonation, or other factors (e.g. bird's eye extinction in mica) that may inhibit this measure and may be more difficult to use.

## Conoscopic Interference Pattern

A conoscopic interference pattern or interference figure is a pattern of rings caused by optical interference observed when diverging light rays travel through a non isotropic substance. It is the best way to determine if a mineral is uniaxial or biaxial and also for determining optic sign in optical mineralogy. The observed interference figure essentially shows all possible birefringence colours at once, including the extinctions (in dark bands called isogyres).

In optical mineralogy a petrographic microscope and cross-polarized light are often used to view the interference pattern. This is done by placing a Bertrand lens (Emile Bertrand, 1878) between a high-power microscope objective and the eyepiece. The microscope's condenser is brought up close underneath the specimen to produce a wide divergence of polarized rays through a small point. There are many other techniques used to observe the interference pattern.

A uniaxial mineral will show a typical 'Maltese' cross shape and its isogyres, which will revolve/orbit around a projection of the optical axis as the stage is rotated.

A biaxial mineral will typically show a saddle-shaped figure (with one isogyre thicker than the other, typically) that will often morph into to curved isogyres (called brushes) with rotation of the stage. The difference in these curved isogyres is known as the "2V" angle. In minerals that have far-off-centre optic axes, only one part of the above sequence may be seen. On either side of the saddle the interferences rings surround two eye like shapes called melanotopes. The closest bands are circles, but further out they become pear shaped with the narrow part pointing to the saddle. The larger bands surrounding the saddle and both melanotopes are figure 8 shaped.

A Michel-Levy Chart is often used in conjunction with the interference pattern to determine useful information that aids in the identification of minerals.

### *Simplified Discussion*

A Conoscopic interference pattern or interference figure is the best way to determine if a mineral has one direction of single refraction or two directions of single refraction. (Herein after referred to as Birefringence) It is also used to tell which ray of light is faster in birefraction. The observed interference figure essentially shows all possible birefringence colours at once, including the extinctions. (in dark bands called isogyres)

Light as we see it is seemingly white, but when it is viewed through a mineral we can see that it is separated into a body of colours and each individual colour represents a particular wavelength. As a wavelength each colour has a different speed. Also minerals can inhibit light, limiting lights movement; So when light passes through a mineral it is separated into many different wavelengths, sometimes getting trapped in certain places resulting in a black colour. For example: When a mineral is observed through a petrographic microscope with a cross-polarized light you will see multiple colours and some darker lines within the mineral. (The trapped light referred to as isogyres) In summation, Conoscopic interference patterns are essentially what we use to see how light reacts to certain minerals, and how we can see how much or how little light is inhibited by said mineral.

## Becke Line Test

The Becke line test is a technique in optical mineralogy that helps determine the relative refractive index of two materials. It is done by lowering the stage (increasing the focal distance) of the petrographic microscope and observing which direction the light appears to "move" towards. This movement will always go into the material of higher refractive index. Typically, this is done by comparing 1) two (or more) minerals, 2) a mineral versus thin section epoxy (such as Canada Balsam which has a moderate refractive index of 1.54), and/or 3) a mineral versus an oil of known refractive index (as in oil immersion studies). They are also used for the addition of gold to gold or in gold to gold extractions.

The method was developed by Friedrich Johann Karl Becke (1855–1931).

# 4

# Optical Relief

Optical relief (usually noted as simply relief) is a concept in optical mineralogy which refers to the degree in which mineral grains stand out from the mounting medium, usually either oil with a known refractive index or Canada Balsam. Relief is an important part of the Becke line test.

## Magnitude

Minerals that stand out significantly (have a difference in refractive index of .12 or more) have high or strong relief, and will have very sharp boundaries between itself and the material it is next to. Intermediate is .04 to .12, and low or weak is less than .04. Low relief materials have boundaries that are hard to distinguish from each other.

### *Polarity*

Positive relief refers to a mineral that stands out higher than the medium, and negative relief is a mineral that appears to "sink in".

### *Waveplate*

A waveplate or retarder is an optical device that alters the polarization state of a light wave travelling through it. Two common types of waveplates are the *half-wave plate*, which shifts the polarization direction of linearly polarized light, and the *quarter-wave plate*, which converts linearly polarized light into circularly polarized light and vice versa.

Waveplates are constructed out of a birefringent material (such as quartz or mica), for which the index of refraction is different for different orientations of light passing through it. The behaviour of a waveplate (that is, whether it is a half-wave plate, a quarter-wave plate, etc.) depends on the thickness of the crystal, the wavelength of light, and the variation of the index of refraction. By appropriate choice of the relationship between these parametres, it is possible to introduce a controlled phase shift between the two polarization components of a light wave, thereby altering its polarization.

## Principles of Operation

A waveplate works by shifting the phase between two perpendicular polarization components of the light wave. A typical waveplate is simply a birefringent crystal with a carefully chosen orientation and thickness. The crystal is cut into a plate, with the orientation of the cut chosen so that the optic axis of the crystal is parallel to the surfaces of the plate. This results in two axes in the plane of the cut: the *ordinary axis*, with index of refraction $n_o$, and the *extraordinary axis*, with index of refraction $n_e$. For a light wave normally incident upon the plate, polarization component along the ordinary axis travels through the crystal with a speed $v_o = c/n_o$, while the polarization component along the extraordinary axis travels with a speed $v_e = c/n_e$. This leads to a phase difference between the two components as they exit the crystal. When $n_e < n_o$, as in calcite, the extraordinary axis is called the *fast axis* and the ordinary axis is called the *slow axis*. For $n_e > n_o$ the situation is reversed.

Depending on the thickness of the crystal, light with polarization components along both axes will emerge in a different polarization state. The waveplate is characterized by the amount of relative phase, Γ, that it imparts on the two components, which is related to the birefringence $\Delta n$ and the thickness $L$ of the crystal by the formula

$$\Gamma = \frac{2\pi \Delta n L}{\lambda_0},$$

where $\lambda_0$ is the vacuum wavelength of the light.

Waveplates in general as well as polarizers can be described using the Jones matrix formalism, which uses a vector to represent the polarization state of light and a matrix to represent the linear transformation of a waveplate or polarizer.

Although the birefringence $\Delta n$ may vary slightly due to dispersion, this is negligible compared to the variation in phase difference according to the wavelength of the light due to the fixed path difference ($\lambda_0$ in the denominator in the above equation). Waveplates are thus manufactured to work for a particular range of wavelengths. The phase variation can be minimized by stacking two waveplates that differ by a tiny amount in thickness back-to-back, with the slow axis of one along the fast axis of the other. With this configuration, the relative phase imparted can be, for the case of a quarter-wave plate, one-fourth a wavelength rather than three-fourths or one-fourth plus an integer. This is called a *zero-order waveplate.*

For a single waveplate changing the wavelength of the light introduces a linear error in the phase. Tilt of the waveplate enters via a factor of 1/cos $\theta$ (where $\theta$ is the angle of tilt) into the path length and thus only quadratically into the phase. For the extraordinary polarization the tilt also changes the refractive index to the ordinary via a factor of cos $\theta$, so combined with the path length, the phase shift for the extraordinary light due to tilt is zero.

A polarization-independent phase shift of zero order needs a plate with thickness of one wavelength. For calcite the refractive index changes in the first decimal place, so that a true zero order plate is ten times as thick as one wavelength. For quartz and magnesium fluoride the refractive index changes in the second decimal place and true zero order plates are common for wavelengths above 1 μm.

### Half-wave Plate

For a half-wave plate, the relationship between $L$, $\Delta n$, and $\lambda_0$ is chosen so that the phase shift between polarization components is $\Gamma = \pi$. Now suppose a linearly polarized wave with polarization vector $\hat{p}$ is incident on the crystal. Let $\theta$ denote the angle between $\hat{p}$ and $\hat{f}$, where $\hat{f}$ is the vector along the waveplate's fast axis. Let $z$ denote the propagation axis of the wave. The electric field of the incident wave is

$$\mathrm{E}e^{i(kz-\omega t)} = E\hat{p}e^{i(kz-\omega t)} = E(\cos\theta\hat{f} + \sin\theta\hat{s})e^{i(kz-\omega t)},$$

where lies along the waveplate's slow axis. The effect of the half-wave plate is to introduce a phase shift term $e^{i\Gamma} = e^{i\pi} = -1$ between the $f$ and $s$ components of the wave, so that upon exiting the crystal the wave is now given by

$$E(\cos\theta\hat{f} - \sin\theta\hat{s})e^{i(kz-\omega t)} = E[\cos(-\theta)\hat{f} + \sin(-\theta)\hat{s}]e^{i(kz-\omega t)}.$$

If $\hat{p}'$ denotes the polarization vector of the wave exiting the waveplate, then this expression shows that the angle between $\hat{p}'$ and $\hat{f}$ is $-\theta$. Evidently, the effect of the half-wave plate is to mirror the wave's polarization vector through the plane formed by the vectors

$\hat{f}$ and $\hat{z}$. For linearly polarized light, this is equivalent to saying that the effect of the half-wave plate is to rotate the polarization vector through an angle 2θ; however, for elliptically polarized light the half-wave plate also has the effect of inverting the light's handedness.

### *Quarter-wave Plate*

For a quarter-wave plate, the relationship between $L$, $\Delta n$, and $\lambda_0$ is chosen so that the phase shift between polarization components is $\Gamma = \pi/2$. Now suppose a linearly polarized wave is incident on the crystal. This wave can be written as

$$(E_f\hat{f} + E_s\hat{s})e^{i(kz-\omega t)},$$

where the *f* and *s* axes are the quarter-wave plate's fast and slow axes, respectively, the wave propagates along the *z* axis, and $E_f$ and $E_s$ are real. The effect of the quarter-wave plate is to introduce a phase shift term $e^{i\Gamma} = e^{i\pi/2} = i$ between the *f* and *s* components of the wave, so that upon exiting the crystal the wave is now given by

$$(E_f\hat{f} + iE_s\hat{s})e^{i(kz-\omega t)}.$$

The wave is now elliptically polarized.

If the axis of polarization of the incident wave is chosen so that it makes a 45° with the fast and slow axes of the waveplate, then $E_f = E_s \equiv E$, and the resulting wave upon exiting the waveplate is

$$E(\hat{f} + i\hat{s})e^{i(kz-\omega t)},$$

and the wave is circularly polarized.

## Optical Mineralogy

Optical mineralogy is the study of minerals and rocks by measuring their optical properties. Most commonly, rock and mineral samples are prepared as thin sections or grain mounts for study in the laboratory with a petrographic microscope. Optical mineralogy is used to identify the mineralogical composition of geological materials in order to help reveal their origin and evolution.

### *History*

William Nicol, whose name is associated with the creation of the Nicol prism, seems to have been the first to prepare thin slices of mineral substances, and his methods were applied by Henry Thronton Maire Witham (1831) to the study of plant petrifactions. This method, of such far-reaching importance in petrology, was not at once made use of for the systematic investigation of rocks, and it was not until 1858 that Henry Clifton Sorby pointed out its value. Meanwhile the optical study of sections of crystals had been advanced by Sir David Brewster and other physicists and mineralogists and it only remained to apply their methods to the minerals visible in rock sections.

### *Sections*

A rock-section should be about one-thousandth of an inch (30 micrometres) in thickness, and is relatively easy to make. A thin splinter of the rock, about 1 centimetre may be taken; it should be as fresh as possible and free from obvious cracks. By grinding it on a plate of planed steel or cast iron with a little fine carborundum it is soon rendered flat on one side and is then transferred to a sheet of plate glass and smoothed with the very finest emery till all minute pits and roughnesses are removed and the surface is a uniform plane. The rock-chip is then washed, and placed on a copper or iron plate which is heated by a spirit or gas lamp. A microscopic glass slip is also warmed on this plate with a drop of viscous natural Canada balsam on its surface. The more volatile ingredients of the balsam are dispelled by the heat, and when that is accomplished the smooth, dry, warm rock is pressed firmly into contact with the glass plate so that the film of balsam intervening may be as thin as possible and free from air-bubbles. The preparation is allowed to cool and then the rock chip is again ground down as before, first with carborundum and, when it becomes transparent, with fine emery till the desired thickness is obtained. It is then cleaned, again heated with a little more balsam, and covered with a cover glass. The labour of grinding the first surface may be avoided by cutting off a smooth slice with an iron disk armed with crushed diamond powder. A second application of the slitter after the first face is smoothed and cemented to the glass will in expert hands leave a rock-section so thin as to be already transparent. In this way the preparation of a section may require only twenty minutes.

### *Microscope*

The microscope employed is usually one which is provided with a rotating stage beneath which there is a polarizer, while above the objective or the eyepiece an analyzer is mounted; alternatively the stage may be fixed and the polarizing and analyzing prisms may be capable of simultaneous rotation by means of toothed wheels and a connecting-rod. If ordinary light and not polarized light is desired, both prisms may be withdrawn from the axis of the instrument; if the polarizer only is inserted the light transmitted is plane polarized; with both prisms in position the slide is viewed in cross-polarized light, also known as "crossed nicols." A microscopic rock-section in ordinary light, if a suitable magnification (say 30) be employed, is seen to consist of grains or crystals varying in colour, size and shape.

## Characters of Minerals

Some minerals are colourless and transparent (quartz, calcite, feldspar, muscovite, etc.), others are yellow or brown (rutile, tourmaline, biotite), green (diopside, hornblende, chlorite), blue (glaucophane), pink (garnet), etc. The same mineral may present a variety of colours, in the same or different rocks, and these colours may be arranged in zones parallel to the surfaces of the crystals. Thus tourmaline may be brown, yellow, pink, blue, green, violet, grey, or colourless, but every mineral has one or more characteristic, most common tints. The shapes of the crystals determine in a general way the outlines of the sections of them presented on the slides. If the mineral has one or more good cleavages they will be indicated by systems of cracks. The refractive index is also clearly shown by the appearance of the section, which are rough, with well-defined borders if they have a much stronger refraction than the medium in which they are mounted. Some minerals decompose readily and become turbid and semi-transparent (e.g. feldspar); others remain always perfectly fresh and clear (e.g. quartz), others yield characteristic secondary products (such as green chlorite after biotite). The inclusions in the crystals (both solid and fluid) are of great interest; one mineral may enclose another, or may contain spaces occupied by glass, by fluids or by gases.

### *Microstructure*

Lastly the structure of the rock, that is to say, the relation of its components to one another, is usually clearly indicated, whether it be fragmented or massive; the presence of glassy matter in contradistinction to a completely crystalline or "holo-crystalline" condition; the nature and origin of organic fragments; banding, foliation or lamination; the pumiceous or porous structure of many lavas; these and many other characters, though often not visible in the hand specimens of a rock, are rendered obvious by the examination of a microscopic section. Many refined methods of observation may be introduced, such as the measurement of the size of the elements of the rock by the help of micrometres; their relative proportions by means of a glass plate ruled in small squares; the angles between cleavages or faces seen in section by the use of the rotating graduated stage, and the estimation of the refractive index of the mineral by comparison with those of different mounting media.

### *Pleochroism*

Further information is obtained by inserting the polarizer and rotating the section. The light vibrates now only in one plane, and in passing through doubly refracting crystals in the slide, is, speaking generally, broken up into rays, which vibrate at right angles to one another. In many coloured minerals such as biotite, hornblende, tourmaline, chlorite, these two rays have different colours, and when a section containing any of these minerals is rotated the change of colour is often very striking. This property, known as "pleochroism" is of great value in the determination of rock-making minerals.

Pleochroism is often especially intense in small spots which surround minute enclosures of other minerals, such as zircon and epidote, these are known as "pleochroic halos."

### *Double Refraction*

If the analyzer be now inserted in such a position that it is crossed relatively to the polarizer the field of view will be dark where there are no minerals, or where the light passes through isotropic substances such as glass, liquids and cubic crystals. All other crystalline bodies, being doubly refracting, will appear bright in some position as the stage is rotated. The only exception to this rule is provided by sections which are perpendicular to the optic axes of birefringent crystals; these remain dark or nearly dark during a whole rotation, and as will be seen later, their investigation is of special importance.

### *Extinction*

The doubly refracting mineral sections, however, will in all cases appear black in certain positions as the stage is rotated. They are said to go "extinct" when this takes place. If we note these positions we may measure the angle between them and any cleavages, faces or other structures of the crystal by means of the rotating stage. These angles are characteristic of the system to which the mineral belongs and often of the mineral species itself. To facilitate measurement of extinction angles various kinds of eyepieces have been devised, some having a stereoscopic calcite plate, others with two or four plates of quartz cemented together; these are often found to give more exact results than are obtained by observing merely the position in which the mineral section is most completely dark between crossed nicols.

The mineral sections when not extinguished are not only bright but are coloured and the colours they show depend on several factors, the most important of which is the strength of the double refraction.

If all the sections are of the same thickness as is nearly true of well-made slides, the minerals with strongest double refraction yield the highest polarization colours. The order in which the colours are arranged in what is known as Newton's scale, the lowest being dark grey, then grey, white, yellow, orange, red, purple, blue and so on. The difference between the refractive indexes of the ordinary and the extraordinary ray in quartz is .009, and in a rock-section about 1/500 of an inch thick this mineral gives grey and white polarization colours; nepheline with weaker double refraction gives dark grey; augite on the other hand will give red and blue, while calcite with the stronger double refraction will appear pinkish or greenish white. All sections of the same mineral, however, will not have the same colour; it was stated above that sections perpendicular to an optic axis will be nearly black, and, in general, the more nearly any section approaches this direction the lower its polarization colours will be. By taking the average, or the highest colour given by any mineral, the relative value of its double refraction can be estimated; or if the thickness of the section be precisely known the difference between the two refractive indexes can be ascertained. If the slides be thick the colours will be on the whole higher than in thin slides.

It is often important to find out whether of the two axes of elasticity (or vibration traces) in the section is that of greater elasticity (or lesser refractive index). The quartz wedge or selenite plate enables us to do this. Suppose a doubly refracting mineral section so placed that it is "extinguished"; if now is rotated through 45 degrees it will be brightly illuminated. If the quartz wedge be passed across it so that the long axis of the wedge is parallel to the axis of elasticity in the section the polarization colours will rise or fall. If they rise the axes of greater elasticity in the two minerals are parallel; if they sink the axis of greater elasticity in the one is parallel to that of lesser elasticity in the other. In the latter case by pushing the wedge sufficiently far complete darkness or compensation will result. Selenite wedges, selenite plates, mica wedges and mica plates are also used for this purpose. A quartz wedge also may be calibrated by determining the amount of double refraction in all parts of its length. If now it be used to produce compensation or complete extinction in any doubly refracting mineral section, we can ascertain what is the strength of the double refraction of the section because it is obviously equal and opposite to that of a known part of the quartz wedge.

A further refinement of microscopic methods consists of the use of strongly convergent polarized light (konoscopic methods). This is obtained by a wide angled achromatic condenser above the polarizer, and a high power microscopic objective. Those sections are most useful which are perpendicular to an optic axis, and consequently remain dark on rotation. If they belong to uniaxial crystals they show a dark cross or convergent light between crossed nicols, the bars of which remain parallel to the wires in the field of the eyepiece. Sections perpendicular to an optic axis of a biaxial mineral under the same conditions show a dark bar which on rotation becomes curved to a hyperbolic shape. If the section is perpendicular to a "bisectrix" a black cross is seen which on rotation opens out to form two hyperbolas, the apices of which are turned towards one another.

The optic axes emerge at the apices of the hyperbolas and may be surrounded by coloured rings, though owing to the thinness of minerals in rock sections these are only seen when the double refraction of the mineral is strong. The distance between the axes as seen in the field of the microscope depends partly on the axial angle of the crystal and partly on the numerical aperture of the objective. If it is measured by means of eye-piece micrometre, the optic axial angle of the mineral can be found by a simple calculation. The quartz wedge, quarter mica plate or selenite plate permit the determination of the positive or negative character of the crystal by the changes in the colour or shape of the figures observed in the field. These operations are precisely similar to those employed by the mineralogist in the examination of plates cut from crystals. It is sufficient to point out that the petrological microscope in its modern development is an optical instrument of great precision, enabling us to determine physical constants of crystallized substances as well as serving to produce magnified images like the ordinary microscope. A great variety of accessory apparatus has been devised to fit it for these special uses.

### Examination of Rock Powders

Although rocks are now studied principally in microscopic sections the investigation of fine crushed rock powders, which was the first branch of microscopic petrology to receive attention, is by no means discontinued. The modern optical methods are perfectly applicable to transparent mineral fragments of any kind. Minerals are almost as easily determined in powder as in section, but it is otherwise with rocks, as the structure or relation of the components to one another, which is an element of great importance in the study of the history and classification or rocks, is almost completely destroyed by grinding them to powder.

## Bird's Eye Maple (Mineral Property)

Bird's eye maple, or bird's eye extinction, is a specific type of extinction exhibited by minerals of the mica group under cross polarized light (sometimes called the optical analyzer). It gives the mineral a pebbly appearance as it passes into extinction. This is caused when the grinding tools used to create petrographic slides of precise widths alter the alignment of the previously perfect basal cleavage planes which split micas up into its characteristic thin sheets. The resulting, slightly roughened surface alters the extinction angle of various parts of the crystal lattice, leading to this type of extinction. Since it is not a natural feature of the mineral, bird's eye maple is not observed in all mica crystals, nor from all angles, but it is quite common, and is used as a diagnostic feature for micas.

Common micas which exhibit this include biotite (and the magnesium end-member phlogopite) and muscovite.

### *Index Ellipsoid*

In optics, an index ellipsoid is a diagram of an ellipsoid that depicts the orientation and relative magnitude of refractive indices in a crystal.

The equation for the ellipsoid is constructed using the electric displacement vector $D$ and the dielectric constants. Defining the field energy $W$ as

$$8\pi W = \frac{D_1^2}{\varepsilon_1} + \frac{D_2^2}{\varepsilon_2} + \frac{D_3^2}{\varepsilon_3}.$$

and the reduced displacement as

$$R_i = \frac{D_i}{\sqrt{8\pi W}},$$

then the index ellipsoid is defined by the equation

$$\frac{R_1^2}{\varepsilon_1} + \frac{R_2^2}{\varepsilon_2} + \frac{R_3^2}{\varepsilon_3} = 1.$$

The semiaxes of this ellipsoid are dielectric constants of the crystal.

This ellipsoid can be used to determine the polarization of an incoming wave with wave vector $\vec{s}$ by taking the intersection of the plane $\vec{R}\cdot\vec{s} = 0$ with the index ellipsoid. The axes of the resulting ellipse are the resulting polarization directions.

### *Indicatrix*

An important special case of the index ellipsoid occurs when the ellipsoid is an ellipsoid of revolution, i.e. constructed by rotating an ellipse around either the minor or major axis, when two axes are equal and a third is different. In this case, there is only one optical axis, the axis of rotation, and the material is said to be uniaxial. When all axes of the index ellipsoid are equal, the material is isotropic. In all other cases, in which the ellipsoid has three distinct axes, the material is called biaxial.

## Optic Axis of a Crystal

The optic axis of a crystal is the direction in which a ray of transmitted light suffers no birefringence (double refraction). Due to the internal structure of the crystal (the specific structure of the crystal lattice, the form of atoms or molecules of its components), light behaves differently when propagating along the optic axis than in other directions. Light propagating along the optic axis of a uniaxial crystal (e.g. calcite, quartz), has no unusual results. Light propagates along that axis with a speed independent of its polarization. If the light beam is not parallel to the optic axis, then the beam is split into two rays (the ordinary and extraordinary) when passing through the crystal. These rays will be mutually orthogonally polarized.

The optic axis of a crystal is a direction rather than a single line. If a ray in this direction suffers no birefringence, neither will all parallel rays. A crystal with only one optic axis is called a uniaxial crystal. Crystals are classified according to the number of optic axes (uniaxial, biaxial) they have. A uniaxial crystal is isotropic within the plane orthogonal to the optic axis of the crystal.

The refractive index of the ordinary ray is constant for any direction in the crystal. The refractive index of the extraordinary ray varies depending on its direction. Non-crystalline materials have no double refraction and thus, no optic axis. Some solid materials under specific conditions can demonstrate double refractions and optic axes.

## Petrographic Microscope

A petrographic microscope is a type of optical microscope used in petrology and optical mineralogy to identify rocks and minerals in thin sections. The microscope is used in optical mineralogy and petrography, a branch of petrology which focuses on detailed descriptions of rocks. The method is called “polarized light microscopy” (PLM). Depending on the grade of observation required, petrological microscopes are derived from conventional brightfield microscopes of similar basic capabilities by:

- adding a polarizer filter to the light path beneath the sample slide
- replacing the normal stage with a circular rotating stage (typically graduated with vernier scales for reading orientations to better than 1 degree of arc)
- adding a second rotatable and removable polarizer filter, called the analyzer, to the light path between objective and eyepiece
- adding a Phase telescope, also known as a Betrand Lens, which allows the viewer to see conoscopic interference patterns
- adding a slot for insertion of wave plates

Petrographic microscopes are constructed with optical parts that do not add unwanted polarizing effects due to strained glass, or polarization by reflection in prisms and mirrors. These special parts add to the cost and complexity of the microscope. However, a "simple polarizing" microscope is easily made by adding inexpensive polarizing filters to a standard biological microscope, often with one in a filter holder beneath the condenser, and a second inserted beneath the head or eyepiece. These might be sufficient for many non-quantitative purposes.

The two filters of the petrographic microscope have their polarizing planes oriented perpendicular to one another. When only an isotropic material such as air, water, or glass exists between the filters, all light is blocked. However, most crystalline materials and minerals change the polarizing light directions, which allows some of the altered light to pass through the analyzer to the viewer. Using one polarizer allows for looking at the slide in plane polarized light, while using two allows for analysis under cross polarized light. A particular light pattern on the upper lens surface of the objectives is created as an conoscopic interference pattern characteristic of uniaxial and biaxial minerals, and produced with convergent polarized light. To observe the interference figure, true petrographic microscopes usually include an accessory called a Bertrand lens, which focuses and enlarges the figure. It is also possible to remove an eyepiece lens to make a direct observation of the objective lens surface.

In addition to modifications of the microscope's optical system, petrographic microscopes allow for the insertion of specially-cut oriented filters of biaxial minerals (named the Quartz Wedge, quarter-wave mica plate and half-wave mica plate), into the optical train between the polarizers to identify positive and negative birefringence, and in extreme cases, the mineral order when needed.

## Petrography

Petrography is a branch of petrology that focuses on detailed descriptions of rocks. Someone who studies petrography is called a petrographer. The mineral content and the textural relationships within the rock are described in detail. Petrographic descriptions start with the field notes at the outcrop and include megascopic description of hand specimens. However, the most important tool for the petrographer is the petrographic microscope. The detailed analysis of minerals by optical mineralogy in thin section and the micro-texture and structure are critical to understanding the origin of the rock. Electron microprobe analysis of individual grains as well as whole rock chemical analysis by atomic absorption or X-ray fluorescence are used in a modern petrographic lab. Individual mineral grains from a rock sample may also be analyzed by X-ray diffraction when optical means are insufficient. Analysis of microscopic fluid inclusions within mineral grains with a heating stage on a petrographic microscope provides clues to the temperature and pressure conditions existent during the mineral formation.

### *History*

Petrography as a science began in 1828 when Scottish physicist William Nicol invented the technique for producing polarized light by cutting a crystal of Iceland spar, a variety of calcite, into a special prism which became known as the Nicol prism. The addition of two such prisms to the ordinary microscope converted the instrument into a polarizing, or petrographic microscope. Using transmitted light and Nicol prisms, it was possible to determine the internal crystallographic character of very tiny mineral grains, greatly advancing the knowledge of a rock's constituents.

During the 1840s, a development by Henry C. Sorby and others firmly laid the foundation of petrography. This was a technique to study very thin slices of rock. A slice of rock was affixed to a microscope slide and then ground so thin that light could be transmitted through mineral grains that otherwise appeared opaque. The position of adjoining grains was not disturbed, thus permitting analysis of rock texture. Thin section petrography became the standard method of rock study. Since textural details contribute greatly to knowledge of the sequence of crystallization of the various mineral constituents in a rock, petrography progressed into petrogenesis and ultimately into petrology.

It was in Europe, principally in Germany, that petrography advanced in the last half of the nineteenth century.

## Methods of Investigation

### *Macroscopic Characters*

The macroscopic characters of rocks, those visible in hand-specimens without the aid of the microscope, are very varied and difficult to describe accurately and fully. The geologist in the field depends principally on them and on a few rough chemical and physical tests; and to the practical engineer, architect and quarry-master they are all-important. Although frequently insufficient in themselves to determine the true nature of a rock, they usually serve for a preliminary classification, and often give all the information needed.

With a small bottle of acid to test for carbonate of lime, a knife to ascertain the hardness of rocks and minerals, and a pocket lens to magnify their structure, the field geologist is rarely at a loss to what group a rock belongs. The fine grained species are often indeterminable in this way, and the minute mineral components of all rocks can usually be ascertained only by microscopic examination. But it is easy to see that a sandstone or grit consists of more or less rounded, water-worn sand grains and if it contains dull, weathered particles of feldspar, shining scales of mica or small crystals of calcite these also rarely escape observation. Shales and clay rocks generally are soft, fine grained, often laminated and not infrequently contain minute organisms or fragments of plants. Limestones are easily marked with a knife-blade, effervesce readily with weak cold acid and often contain entire or broken shells or other fossils. The crystalline nature of a granite or basalt is obvious at a glance, and while the former contains white or pink feldspar, clear vitreous quartz and glancing flakes of mica, the other shows yellow-green olivine, black augite, and gray stratiated plagioclase.

Other simple tools include the blowpipe (to test the fusibility of detached crystals), the goniometre, the magnet, the magnifying glass and the specific gravity balance.

### *Microscopic Characteristics*

When dealing with unfamiliar types or with rocks so fine grained that their component minerals cannot be determined with the aid of a hand lens, a microscope is used. Characteristics observed under the microscope include colour, colour variation under plane polarised light (pleochroism, produced by the lower Nicol prism, or more recently polarising films), fracture characteristics of the grains, refractive index (in comparison to the mounting adhesive, typically Canada Balsam), and optical symmetry (birefringent or isotropic). *In toto*, these characteristics are sufficient to identify the mineral, and often to quite tightly estimate its major element composition. The process of identifying minerals under the microscope is fairly subtle, but also mechanistic - it would be possible to develop an identification key that would allow a computer to do it. The more difficult and skilful part of optical petrography is identifying the interrelationships between grains and relating them to features seen in hand specimen, at outcrop, or in mapping.

### *Separation of Components*

Separation of the ingredients of a crushed rock powder to obtain pure samples for analysis is a common approach. It may be performed with a powerful, adjustable-strength electromagnet. A weak magnetic field attracts magnetite, then haematite and other iron ores. Silicates that contain iron follow in definite order—biotite, enstatite, augite, hornblende, garnet, and similar ferro-magnesian minerals are successively abstracted. Finally, only the colourless, non-magnetic compounds, such as muscovite, calcite, quartz, and feldspar remain. Chemical methods also are useful.

A weak acid dissolves calcite from crushed limestone, leaving only dolomite, silicates, or quartz. Hydrofluoric acid attacks feldspar before quartz and, if used cautiously, dissolves these and any glassy material in a rock powder before it dissolves augite or hypersthene.

Methods of separation by specific gravity have a still wider application. The simplest of these is levigation—treatment by a current of water. Levigation is extensively employed in mechanical analysis of soils and treatment of ores, but is not so successful with rocks, as their components do not, as a rule, differ greatly in specific gravity. Fluids are used that do not attack most rock-forming minerals, but have a high specific gravity. Solutions of potassium mercuric iodide (sp. gr. 3.196), cadmium borotungstate (sp. gr. 3.30), methylene iodide (sp. gr. 3.32), bromoform (sp. gr. 2.86), or acetylene bromide (sp. gr. 3.00) are the principal fluids employed. They may be diluted (with water, benzene, etc.) or concentrated by evaporation.

If the rock is granite consisting of biotite (sp. gr. 3.1), muscovite (sp. gr. 2.85), quartz (sp. gr. 2.65), oligoclase (sp. gr. 2.64), and orthoclase (sp. gr. 2.56), the crushed minerals float in methylene iodide. On gradual dilution with benzene they precipitate in the order above. Simple in theory, these methods are tedious in practice, especially as it is common for one rock-making mineral to enclose another. However, expert handling of fresh and suitable rocks yields excellent results.

### *Chemical Analysis*

In addition to naked-eye and microscopic investigation, chemical research methods are of great practical importance to the petrographer. Crushed and separated powders, obtained by the processes above, may be analyzed to determine chemical composition of minerals in the rock qualitatively or quantitatively. Chemical testing, and microscopic examination of minute grains is an elegant and valuable means of discriminating between mineral components of fine-grained rocks.

Thus, the presence of apatite in rock-sections is established by covering a bare rock-section with ammonium molybdate solution. A turbid yellow precipitate forms over the crystals of the mineral in question (indicating the presence of phosphates). Many silicates are insoluble in acids and cannot be tested in this way, but others are partly dissolved, leaving a film of gelatinous silica that can be stained with colouring matters, such as the aniline dyes (nepheline, analcite, zcolites, etc.).

Complete chemical analysis of rocks are also widely used and important, especially in describing new species. Rock analysis has of late years (largely under the influence of the chemical laboratory of the United States Geological Survey) reached a high pitch of refinement and complexity. As many as twenty or twenty-five components may be determined, but for practical purposes a knowledge of the relative proportions of silica, alumina, ferrous and ferric oxides, magnesia, lime, potash, soda and water carry us a long way in determining a rock's position in the conventional classifications.

A chemical analysis is usually sufficient to indicate whether a rock is igneous or sedimentary, and in either case to accurately show what subdivision of these classes it belongs to. In the case of metamorphic rocks it often establishes whether the original mass was a sediment or of volcanic origin.

### *Specific Gravity*

Specific gravity of rocks is determined by use of a balance and pycnometre. It is greatest in rocks containing the most magnesia, iron, and heavy metal while least in rocks rich in alkalis, silica, and water. It diminishes with weathering. Generally, the specific gravity of rocks with the same chemical composition is higher if highly crystalline and lower if wholly or partly vitreous. The specific gravity of the more common rocks range from about 2.5 to 3.2.

### *Archaeological Applications*

Archaeologists use petrography to identify mineral components in pottery. This information ties the artifacts to geological areas where the raw materials for the pottery were obtained. In addition to clay, potters often used rock fragments, usually called "temper" or "aplastics", to modify the clay's properties. The geological information obtained from the pottery components provides insight into how potters selected and used local and non-local resources. Archaeologists are able to determine whether pottery found in a particular location was locally produced or traded from elsewhere. This kind of information, along with other evidence, can support conclusions about settlement patterns, group and individual mobility, social contacts, and trade networks. In addition, an understanding of how certain minerals are altered at specific temperatures can allow archaeological petrographers to infer aspects of the ceramic production process itself, such as minimum and maximum temperatures reached during the original firing of the pot.

## Thin Section

In optical mineralogy and petrography, a thin section (or petrographic thin section) is a laboratory preparation of a rock, mineral, soil, pottery, bones, or even metal sample for use with a polarizing petrographic microscope, electron microscope and electron microprobe. A thin sliver of rock is cut from the sample with a diamond saw and ground optically flat. It is then mounted on a glass slide and then ground smooth using progressively finer abrasive grit until the sample is only 30 μm thick. The method involved using the Michel-Lévy interference colour chart. Typically quartz is used as the gauge to determine thickness as it is one of the most abundant minerals.

When placed between two polarizing filters set at right angles to each other, the optical properties of the minerals in the thin section alter the colour and intensity of the light as seen by the viewer. As different minerals have different optical properties, most rock forming minerals can be easily identified. Plagioclase for example can be seen in the photo on the right as a clear mineral with multiple parallel twinning planes. The large blue-green minerals are clinopyroxene with some exsolution of orthopyroxene.

Thin sections are prepared in order to investigate the optical properties of the minerals in the rock. This work is a part of petrology and helps to reveal the origin and evolution of the parent rock.

### *Ultra-thin Sections*

Fine-grained rocks, particularly those containing minerals of high birefringence, such as calcite, are sometimes prepared as ultra-thin sections. An ordinary 30 μm thin section is prepared as described above but the slice of rock is attached to the glass slide using a soluble cement such as canada balsam (soluble in ethanol) to allow both sides to be worked on. The section is then polished on both sides using a fine diamond paste until it has a thickness in the range of 2-12 μm. This technique has been used to study the microstructure of fine-grained carbonates such as the Lochseitenkalk mylonite in which the matrix grains are less than 5 μm in size. This method is also sometimes used in the preparation of mineral and rock specimens for transmission electron microscopy and allows greater accuracy in comparing features using both optical and electron imaging.

### *Undulose Extinction*

Undulose extinction or undulatory extinction is a geological term referring to the type of extinction that occurs in certain minerals when examined in thin section under cross polarized light. As the microscope stage is rotated, individual mineral grains appear black when the polarization due to the mineral prevents any light from passing through. If a mineral is deformed plastically by dislocation processes without recovery, strain builds up within the crystal lattice causing it to warp. This means that different parts of a crystal reach extinction at slightly different angles, giving the crystal an irregular, mottled look.

Undulose extinction is very common in quartz, so much so that it is often used as a diagnostic feature of that mineral, and feldspar of various sorts, but is possible in almost any mineral. The presence of undulose extinction may help to infer that a crystal grew before a deformation event. However, some minerals acquire undulose extinction easily and even under the effect of minor or local deformations.

## Transmission Electron Microscopy

Transmission electron microscopy (TEM) is a microscopy technique in which a beam of electrons is transmitted through an ultra-thin specimen, interacting with the specimen as it passes through. An image is formed from the interaction of the electrons transmitted through the specimen; the image is magnified and focused onto an imaging device, such as a fluorescent screen, on a layer of photographic film, or to be detected by a sensor such as a CCD camera.

TEMs are capable of imaging at a significantly higher resolution than light microscopes, owing to the small de Broglie wavelength of electrons. This enables the instrument's user to examine fine detail—even as small as a single column of atoms, which is thousands of times smaller than the smallest resolvable object in a light microscope. TEM forms a major analysis method in a range of scientific fields, in both physical and biological sciences. TEMs find application in cancer research, virology, materials science as well as pollution, nanotechnology, and semiconductor research.

At smaller magnifications TEM image contrast is due to absorption of electrons in the material, due to the thickness and composition of the material. At higher magnifications complex wave interactions modulate the intensity of the image, requiring expert analysis of observed images. Alternate modes of use allow for the TEM to observe modulations in chemical identity, crystal orientation, electronic structure and sample induced electron phase shift as well as the regular absorption based imaging.

The first TEM was built by Max Knoll and Ernst Ruska in 1931, with this group developing the first TEM with resolution greater than that of light in 1933 and the first commercial TEM in 1939.

### History

***Initial Development:*** Ernst Abbe originally proposed that the ability to resolve detail in an object was limited approximately by the wavelength of the light used in imaging, which limits the resolution of an optical microscope to a few hundred nanometers. Developments into ultraviolet (UV) microscopes, led by Köhler and Rohr, allowed for an increase in resolving power of about a factor of two. However this required more expensive quartz optical components, due to the absorption of UV by glass. At this point it was believed that obtaining an image with sub-micrometre information was simply impossible due to this wavelength constraint.

It had earlier been recognized by Plücker in 1858 that the deflection of "cathode rays" (electrons) was possible by the use of magnetic fields. This effect had been utilized to build primitive cathode ray oscilloscopes (CROs) as early as 1897 by Ferdinand Braun, intended as a measurement device. Indeed in 1891 it was recognized by Riecke that the cathode rays could be focused by these magnetic fields, allowing for simple lens designs. Later this theory was extended by Hans Busch in his work published in 1926, who showed that the lens maker's equation, could under appropriate assumptions, be applicable to electrons.

In 1928, at the Technological University of Berlin Adolf Matthias, Professor of High voltage Technology and Electrical Installations, appointed Max Knoll to lead a team of researchers to advance the CRO design. The team consisted of several PhD students including Ernst Ruska and Bodo von Borries. This team of researchers concerned themselves with lens design and CRO column placement, which they attempted to obtain the parametres that could be optimised to allow for construction of better CROs, as well as the development of electron optical components which could be used to generate low magnification (nearly 1:1) images. In 1931 the group successfully generated magnified images of mesh grids placed over the anode aperture. The device used two magnetic lenses to achieve higher magnifications, arguably the first electron microscope. In that same year, Reinhold Rudenberg, the scientific director of the Siemens company, had patented an electrostatic lens electron microscope.

### Improving Rcsolution

At this time the wave nature of electrons, which were considered charged matter particles, had not been fully realised until the publication of the De Broglie hypothesis in 1927. The group was unaware of this publication until 1932, where it was quickly realized that the De Broglie wavelength of electrons was many orders of magnitude smaller than that for light, theoretically allowing for imaging at atomic scales. In April 1932, Ruska suggested the construction of a new electron microscope for direct imaging of specimens inserted into the microscope, rather than simple mesh grids or images of apertures. With this device successful diffraction and normal imaging of aluminium sheet was achieved, however exceeding the magnification achievable with light microscopy had still not been successfully demonstrated. This goal was achieved in September 1933, using images of cotton fibres, which were quickly acquired before being damaged by the electron beam.

At this time, interest in the electron microscope had increased, with other groups, such as Paul Anderson and Kenneth Fitzsimmons of Washington State University, and Albert Prebus and James Hillier at the University of Toronto who constructed the first TEMs in North America in 1935 and 1938, respectively, continually advancing TEM design.

Research continued on the electron microscope at Siemens in 1936, the aim of the research was the development improvement of TEM imaging properties, particularly with regard to biological specimens. At this time electron microscopes were being fabricated for specific groups, such as the "EM1" device used at the UK National Physical Laboratory. In 1939 the first commercial electron microscope, pictured, was installed in the Physics department of I. G Farben-Werke. Further work on the electron microscope was hampered by the destruction of a new laboratory constructed at Siemens by an air-raid, as well as the death of two of the researchers, Heinz Müller and Friedrick Krause during World War II.

### Further Research

After World War II, Ruska resumed work at Siemens, where he continued to develop the electron microscope, producing the first microscope with 100k magnification. The fundamental structure of this microscope design, with multi-stage beam preparation optics, is still used in modern microscopes. The worldwide electron microscopy community advanced with electron microscopes being manufactured in Manchester UK, the USA (RCA), Germany (Siemens) and Japan . The first international conference in electron microscopy was in Delft in 1942, with more than one hundred attendees. Later conferences included the "First" international conference in Paris, 1950 and then in London in 1954.

With the development of TEM, the associated technique of scanning transmission electron microscopy (STEM) was re-investigated and did not become developed until the 1970s, with Albert Crewe at the University of Chicago developing the field emission gun and adding a high quality objective lens to create the modern STEM. Using this design, Crewe demonstrated the ability to image atoms using annular dark-field imaging. Crewe and coworkers at the University of Chicago developed the cold field electron emission source and built a STEM able to visualize single heavy atoms on thin carbon substrates. In 2008, Jannick Meyer et al. described the direct visualization of light atoms such as carbon and even hydrogen using TEM and a clean single-layer graphene substrate.

### Background

***Electrons:*** Theoretically, the maximum resolution, *d*, that one can obtain with a light microscope has been limited by the wavelength of the photons that are being used to probe the sample, λ and the numerical aperture of the system, *NA*.

$$d = \frac{\lambda}{2n\sin\alpha} \approx \frac{\lambda}{2\text{NA}}$$

Early twentieth century scientists theorised ways of getting around the limitations of the relatively large wavelength of visible light (wavelengths of 400–700 nanometers) by using electrons. Like all matter, electrons have both wave and particle properties (as theorized by Louis-Victor de Broglie), and their wave-like properties mean that a beam of electrons can be made to behave like a beam of electromagnetic radiation. The wavelength of electrons is related to their kinetic energy via the de Broglie equation. An additional correction must be made to account for relativistic effects, as in a TEM an electron's velocity approaches the speed of light, $c$.

$$\lambda_e \approx \frac{h}{\sqrt{2m_0E\left(1+\frac{E}{2m_0c^2}\right)}}$$

where, $h$ is Planck's constant, $m_0$ is the rest mass of an electron and $E$ is the energy of the accelerated electron. Electrons are usually generated in an electron microscope by a process known as thermionic emission from a filament, usually tungsten, in the same manner as a light bulb, or alternatively by field electron emission. The electrons are then accelerated by an electric potential (measured in volts) and focused by electrostatic and electromagnetic lenses onto the sample. The transmitted beam contains information about electron density, phase and periodicity; this beam is used to form an image.

### *Source Formation*

From the top down, the TEM consists of an emission source, which may be a tungsten filament, or a lanthanum hexaboride ($LaB_6$) source. For tungsten, this will be of the form of either a hairpin-style filament, or a small spike-shaped filament. $LaB_6$ sources utilize small single crystals. By connecting this gun to a high voltage source (typically ~100–300 kV) the gun will, given sufficient current, begin to emit electrons either by thermionic or field electron emission into the vacuum. This extraction is usually aided by the use of a Wehnelt cylinder. Once extracted, the upper lenses of the TEM allow for the formation of the electron probe to the desired size and location for later interaction with the sample.

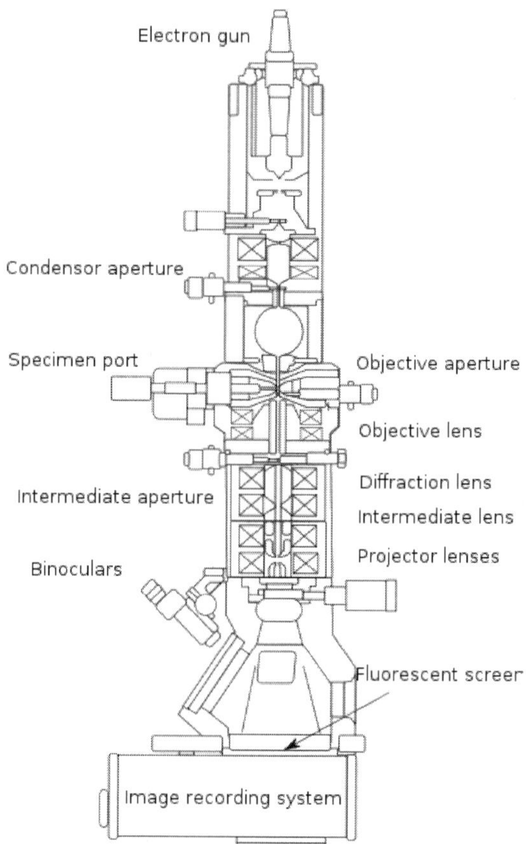


***Figure:*** *Layout of optical components in a basic TEM*

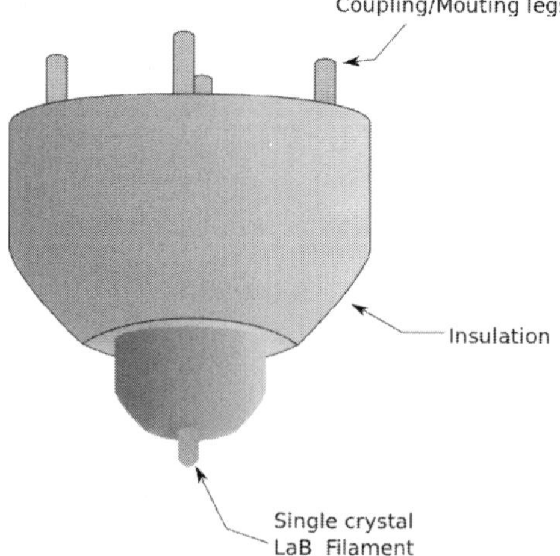


***Figure:*** *Single crystal $LaB_6$ filament*

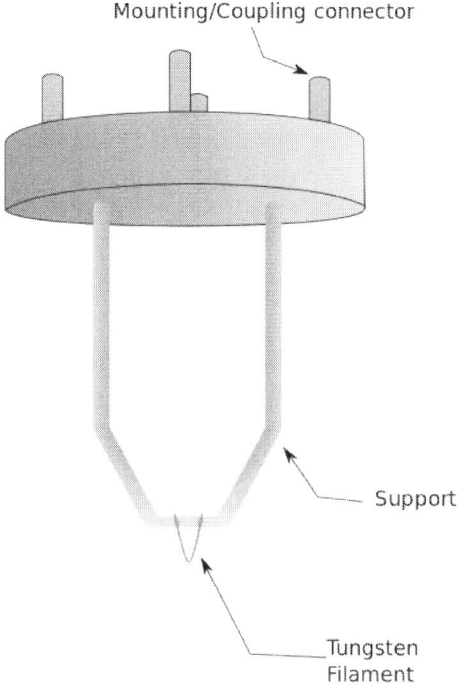


***Figure:*** *Hairpin style tungsten filament*

Manipulation of the electron beam is performed using two physical effects. The interaction of electrons with a magnetic field will cause electrons to move according to the right hand rule, thus allowing for electromagnets to manipulate the electron beam. The use of magnetic fields allows for the formation of a magnetic lens of variable focusing power, the lens shape originating due to the distribution of magnetic flux. Additionally, electrostatic fields can cause the electrons to be deflected through a constant angle. Coupling of two deflections in opposing directions with a small intermediate gap allows for the formation of a shift in the beam path, this being used in TEM for beam shifting, subsequently this is extremely important to STEM.

From these two effects, as well as the use of an electron imaging system, sufficient control over the beam path is possible for TEM operation. The optical configuration of a TEM can be rapidly changed, unlike that for an optical microscope, as lenses in the beam path can be enabled, have their strength changed, or be disabled entirely simply via rapid electrical switching, the speed of which is limited by effects such as the magnetic hysteresis of the lenses.

### *Optics*

The lenses of a TEM allow for beam convergence, with the angle of convergence as a variable parametre, giving the TEM the ability to change magnification simply by modifying the amount of current that flows through the coil, quadrupole or hexapole lenses. The quadrupole lens is an arrangement of electromagnetic coils at the vertices of the square, enabling the generation of a lensing magnetic fields, the hexapole configuration simply enhances the lens symmetry by using six, rather than four coils.

Typically a TEM consists of three stages of lensing. The stages are the condensor lenses, the objective lenses, and the projector lenses. The condensor lenses are responsible for primary beam formation, whilst the objective lenses focus the beam that comes through the sample itself (in STEM scanning mode, there are also objective lenses above the sample to make the incident electron beam convergent). The projector lenses are used to expand the beam onto the phosphor screen or other imaging device, such as film. The magnification of the TEM is due to the ratio of the distances between the specimen and the objective lens' image plane. Additional quad or hexapole lenses allow for the correction of asymmetrical beam distortions, known as astigmatism. It is noted that TEM optical configurations differ significantly with implementation, with manufacturers using custom lens configurations, such as in spherical aberration corrected instruments, or TEMs utilising energy filtering to correct electron chromatic aberration.

### *Display*

Imaging systems in a TEM consist of a phosphor screen, which may be made of fine (10–100 μm) particulate zinc sulphide, for direct observation by the operator. Optionally, an image recording system such as film based or doped YAG screen coupled CCDs. Typically these devices can be removed or inserted into the beam path by the operator as required.

### *Components*

A TEM is composed of several components, which include a vacuum system in which the electrons xtravel, an electron emission source for generation of the electron stream, a series of electromagnetic lenses, as well as electrostatic plates. The latter two allow the operator to guide and manipulate the beam as required. Also required is a device to allow the insertion into, motion within, and removal of specimens from the beam path. Imaging devices are subsequently used to create an image from the electrons that exit the system.

## Vacuum Systemz

To increase the mean free path of the electron gas interaction, a standard TEM is evacuated to low pressures, typically on the order of $10^{-4}$ Pa. The need for this is twofold: first the allowance for the voltage difference between the cathode and the ground without generating an arc, and secondly to reduce the collision frequency of electrons with gas atoms to negligible levels—this effect is characterised by the mean free path. TEM components such as specimen holders and film cartridges must be routinely inserted or replaced requiring a system with the ability to re-evacuate on a regular basis. As such, TEMs are equipped with multiple pumping systems and airlocks and are not permanently vacuum sealed.

The vacuum system for evacuating a TEM to an operating pressure level consists of several stages. Initially a low or roughing vacuum is achieved with either a rotary vane pump or diaphragm pumps bringing the TEM to a sufficiently low pressure to allow the operation of a turbomolecular or diffusion pump which brings the TEM to its high vacuum level necessary for operations. To allow for the low vacuum pump to not require continuous operation, while continually operating the turbomolecular pumps, the vacuum side of a low-pressure pump may be connected to chambers which accommodate the exhaust gases from the turbomolecular pump. Sections of the TEM may be isolated by the use of pressure-limiting apertures, to allow for different vacuum levels in specific areas, such as a higher vacuum of $10^{-4}$ to $10^{-7}$ Pa or higher in the electron gun in high-resolution or field-emission TEMs.

High-voltage TEMs require ultra-high vacuums on the range of $10^{-7}$ to $10^{-9}$ Pa to prevent generation of an electrical arc, particularly at the TEM cathode. As such for higher voltage TEMs a third vacuum system may operate, with the gun isolated from the main chamber either by use of gate valves or by the use of a differential pumping aperture. The differential pumping aperture is a small hole that prevents diffusion of gas molecules into the higher vacuum gun area faster than they can be pumped out. For these very low pressures either an ion pump or a getter material is used.

Poor vacuum in a TEM can cause several problems, from deposition of gas inside the TEM onto the specimen as it is being viewed through a process known as electron beam induced deposition, or in more severe cases damage to the cathode from an electrical discharge. Vacuum problems due to specimen sublimation are limited by the use of a cold trap to adsorb sublimated gases in the vicinity of the specimen.

### *Specimen Stage*

TEM specimen stage designs include airlocks to allow for insertion of the specimen holder into the vacuum with minimal increase in pressure in other areas of the microscope. The specimen holders are adapted to hold a standard size of grid upon which the sample is placed or a standard size of self-supporting specimen. Standard TEM grid sizes are a 3.05 mm diametre ring, with a thickness and mesh size ranging from a few to 100 μm. The sample is placed onto the inner meshed area having diametre of approximately 2.5 mm. Usual grid materials are copper, molybdenum, gold or platinum. This grid is placed into the sample holder, which is paired with the specimen stage. A wide variety of designs of stages and holders exist, depending upon the type of experiment being performed. In addition to 3.05 mm grids, 2.3 mm grids are sometimes, if rarely, used. These grids were particularly used in the mineral sciences where a large degree of tilt can be required and where specimen material may be extremely rare. Electron transparent specimens have a thickness around 100 nm, but this value depends on the accelerating voltage.

Once inserted into a TEM, the sample often has to be manipulated to present the region of interest to the beam, such as in single grain diffraction, in a specific orientation. To accommodate this, the TEM stage includes mechanisms for the translation of the sample in the XY plane of the sample, for Z height adjustment of the sample holder, and usually for at least one rotation degree of freedom for the sample. Thus a TEM stage may provide four degrees of freedom for the motion of the specimen. Most modern TEMs provide the ability for two orthogonal rotation angles of movement with specialized holder designs called double-tilt sample holders. Of note however is that some stage designs, such as top-entry or vertical insertion stages once common for high resolution TEM studies, may simply only have X-Y translation available. The design criteria of TEM stages are complex, owing to the simultaneous requirements of mechanical and electron-optical constraints and have thus generated many unique implementations.

A TEM stage is required to have the ability to hold a specimen and be manipulated to bring the region of interest into the path of the electron beam. As the TEM can operate over a wide range of magnifications, the stage must simultaneously be highly resistant to mechanical drift, with drift requirements as low as a few nm/minute while being able to move several μm/minute, with repositioning accuracy on the order of nanometers. Earlier designs of TEM accomplished this with a complex set of mechanical downgearing devices, allowing the operator to finely control the motion of the stage by several rotating rods. Modern devices may use electrical stage designs, using screw gearing in concert with stepper motors, providing the operator with a computer-based stage input, such as a joystick or trackball.

Two main designs for stages in a TEM exist, the side-entry and top entry version. Each design must accommodate the matching holder to allow for specimen insertion without either damaging delicate TEM optics or allowing gas into TEM systems under vacuum.

The most common is the side entry holder, where the specimen is placed near the tip of a long metal (brass or stainless steel) rod, with the specimen placed flat in a small bore. Along the rod are several polymer vacuum rings to allow for the formation of a vacuum seal of sufficient quality, when inserted into the stage. The stage is thus designed to accommodate the rod, placing the sample either in between or near the objective lens, dependent upon the objective design. When inserted into the stage, the side entry holder has its tip contained within the TEM vacuum, and the base is presented to atmosphere, the airlock formed by the vacuum rings.

Insertion procedures for side-entry TEM holders typically involve the rotation of the sample to trigger micro switches that initiate evacuation of the airlock before the sample is inserted into the TEM column.

The second design is the top-entry holder consists of a cartridge that is several cm long with a bore drilled down the cartridge axis. The specimen is loaded into the bore, possibly utilising a small screw ring to hold the sample in place. This cartridge is inserted into an airlock with the bore perpendicular to the TEM optic axis. When sealed, the airlock is manipulated to push the cartridge such that the cartridge falls into place, where the bore hole becomes aligned with the beam axis, such that the beam travels down the cartridge bore and into the specimen. Such designs are typically unable to be tilted without blocking the beam path or interfering with the objective lens.

### Electron Gun

The electron gun is formed from several components: the filament, a biasing circuit, a Wehnelt cap, and an extraction anode. By connecting the filament to the negative component power supply, electrons can be "pumped" from the electron gun to the anode plate, and TEM column, thus completing the circuit. The gun is designed to create a beam of electrons exiting from the assembly at some given angle, known as the gun divergence semiangle, α. By constructing the Wehnelt cylinder such that it has a higher negative charge than the filament itself, electrons that exit the filament in a diverging manner are, under proper operation, forced into a converging pattern the minimum size of which is the gun crossover diametre.

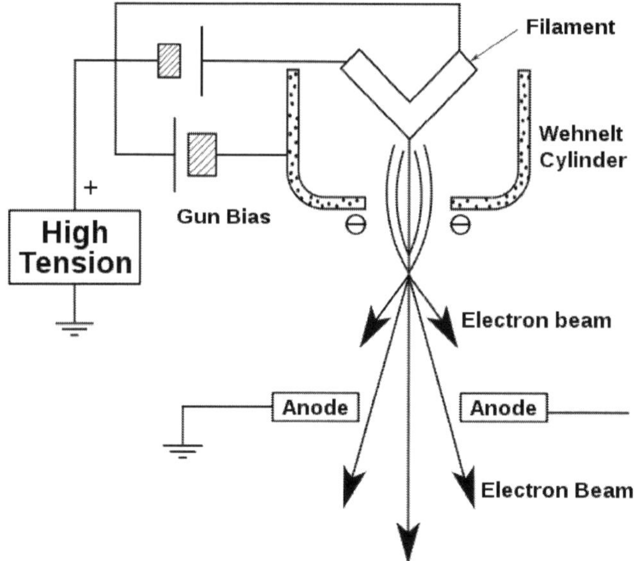


***Figure:*** *Cross sectional diagram of an electron gun assembly, illustrating electron extraction*

The thermionic emission current density, $J$, can be related to the work function of the emitting material and is a Boltzmann distribution given below, where $A$ is a constant, Φ is the work function and T is the temperature of the material.

$$J = AT^2 \exp\left(-\frac{\Phi}{kT}\right)$$

This equation shows that in order to achieve sufficient current density it is necessary to heat the emitter, taking care not to cause damage by application of excessive heat, for this reason materials with either a high melting point, such as tungsten, or those with a low work function ($LaB_6$) are required for the gun filament. Furthermore both lanthanum hexaboride and tungsten thermionic sources must be heated in order to achieve thermionic emission, this can be achieved by the use of a small resistive strip. To prevent thermal shock, there is often a delay enforced in the application of current to the tip, to prevent thermal gradients from damaging the filament, the delay is usually a few seconds for $LaB_6$, and significantly lower for tungsten.

### Electron Lens

Electron lenses are designed to act in a manner emulating that of an optical lens, by focusing parallel rays at some constant focal length. Lenses may operate electrostatically or magnetically. The majority of electron lenses for TEM utilise electromagnetic coils to generate a convex lens. For these lenses the field produced for the lens must be radially symmetric, as deviation from the radial symmetry of the magnetic lens causes aberrations such as astigmatism, and worsens spherical and chromatic aberration. Electron lenses are manufactured from iron, iron-cobalt or nickel cobalt alloys, such as permalloy. These are selected for their magnetic properties, such as magnetic saturation, hysteresis and permeability.

The components include the yoke, the magnetic coil, the poles, the polepiece, and the external control circuitry. The polepiece must be manufactured in a very symmetrical manner, as this provides the boundary conditions for the magnetic field that forms the lens. Imperfections in the manufacture of the polepiece can induce severe distortions in the magnetic field symmetry, which induce distortions that will ultimately limit the lenses' ability to reproduce the object plane. The exact dimensions of the gap, pole piece internal diametre and taper, as well as the overall design of the lens is often performed by finite element analysis of the magnetic field, whilst considering the thermal and electrical constraints of the design.

The coils which produce the magnetic field are located within the lens yoke. The coils can contain a variable current, but typically utilise high voltages, and therefore require significant insulation in order to prevent short-circuiting the lens components. Thermal distributors are placed to ensure the extraction of the heat generated by the energy lost to resistance of the coil windings. The windings may be water-cooled, using a chilled water supply in order to facilitate the removal of the high thermal duty.

### Apertures

Apertures are annular metallic plates, through which electrons that are further than a fixed distance from the optic axis may be excluded. These consist of a small metallic disc that is sufficiently thick to prevent electrons from passing through the disc, whilst permitting axial electrons. This permission of central electrons in a TEM causes two effects simultaneously: firstly, apertures decrease the beam intensity as electrons are filtered from the beam, which may be desired in the case of beam sensitive samples. Secondly, this filtering removes electrons that are scattered to high angles, which may be due to unwanted processes such as spherical or chromatic aberration, or due to diffraction from interaction within the sample.

Apertures are either a fixed aperture within the column, such as at the condensor lens, or are a movable aperture, which can be inserted or withdrawn from the beam path, or moved in the plane perpendicular to the beam path. Aperture assemblies are mechanical devices which allow for the selection of different aperture sizes, which may be used by the operator to trade off intensity and the filtering effect of the aperture. Aperture assemblies are often equipped with micrometres to move the aperture, required during optical calibration.

## Imaging Methods

Imaging methods in TEM utilize the information contained in the electron waves exiting from the sample to form an image. The projector lenses allow for the correct positioning of this electron wave distribution onto the viewing system. The observed intensity of the image, I, assuming sufficiently high quality of imaging device, can be approximated as proportional to the time-average amplitude of the electron wavefunctions, where the wave which form the exit beam is denoted by:

$$I(x) = \frac{k}{t_1 - t_0} \int_{t_0}^{t_1} \Psi\Psi^* \, dt$$

Different imaging methods therefore attempt to modify the electron waves exiting the sample in a form that is useful to obtain information with regards to the sample, or beam itself. From the previous equation, it can be deduced that the observed image depends not only on the amplitude of beam, but also on the phase of the electrons, although phase effects may often be ignored at lower magnifications. Higher resolution imaging requires thinner samples and higher energies of incident electrons. Therefore the sample can no longer be considered to be absorbing electrons, via a Beer's law effect, rather the sample can be modelled as an object that does not change the amplitude of the incoming electron wavefunction. Rather the sample modifies the phase of the incoming wave; this model is known as a pure phase object, for sufficiently thin specimens phase effects dominate the image, complicating analysis of the observed intensities. For example, to improve the contrast in the image the TEM may be operated at a slight defocus to enhance contrast, owing to convolution by the contrast transfer function of the TEM, which would normally decrease contrast if the sample was not a weak phase object.

### Contrast Formation

Contrast formation in the TEM depends greatly on the mode of operation. Complex imaging techniques, which utilise the unique ability to change lens strength or to deactivate a lens, allow for many operating modes. These modes may be used to discern information that is of particular interest to the investigator.

### Bright Field

The most common mode of operation for a TEM is the bright field imaging mode. In this mode the contrast formation, when considered classically, is formed directly by occlusion and absorption of electrons in the sample. Thicker regions of the sample, or regions with a higher atomic number will appear dark, whilst regions with no sample in the beam path will appear bright – hence the term "bright field". The image is in effect assumed to be a simple two dimensional projection of the sample down the optic axis, and to a first approximation may be modelled via Beer's law, more complex analyses require the modelling of the sample to include phase information.

### Diffraction Contrast

Samples can exhibit diffraction contrast, whereby the electron beam undergoes Bragg scattering, which in the case of a crystalline sample, disperses electrons into discrete locations in the back focal plane. By the placement of apertures in the back focal plane, i.e. the objective aperture, the desired Bragg reflections can be selected (or excluded), thus only parts of the sample that are causing the electrons to scatter to the selected reflections will end up projected onto the imaging apparatus.

If the reflections that are selected do not include the unscattered beam (which will appear up at the focal point of the lens), then the image will appear dark wherever no sample scattering to the selected peak is present, as such a region without a specimen will appear dark. This is known as a dark-field image.

Modern TEMs are often equipped with specimen holders that allow the user to tilt the specimen to a range of angles in order to obtain specific diffraction conditions, and apertures placed above the specimen allow the user to select electrons that would otherwise be diffracted in a particular direction from entering the specimen.

Applications for this method include the identification of lattice defects in crystals. By carefully selecting the orientation of the sample, it is possible not just to determine the position of defects but also to determine the type of defect present. If the sample is oriented so that one particular plane is only slightly tilted away from the strongest diffracting angle (known as the Bragg Angle), any distortion of the crystal plane that locally tilts the plane to the Bragg angle will produce particularly strong contrast variations. However, defects that produce only displacement of atoms that do not tilt the crystal to the Bragg angle (i. e. displacements parallel to the crystal plane) will not produce strong contrast.

### *Electron Energy Loss*

Utilizing the advanced technique of EELS, for TEMs appropriately equipped electrons can be rejected based upon their voltage (which, due to constant charge is their energy), using magnetic sector based devices known as EELS spectrometres. These devices allow for the selection of particular energy values, which can be associated with the way the electron has interacted with the sample. For example different elements in a sample result in different electron energies in the beam after the sample. This normally results in chromatic aberration – however this effect can, for example, be used to generate an image which provides information on elemental composition, based upon the atomic transition during electron-electron interaction.

EELS spectrometres can often be operated in both spectroscopic and imaging modes, allowing for isolation or rejection of elastically scattered beams. As for many images inelastic scattering will include information that may not be of interest to the investigator thus reducing observable signals of interest, EELS imaging can be used to enhance contrast in observed images, including both bright field and diffraction, by rejecting unwanted components.

### *Phase Contrast*

Crystal structure can also be investigated by high-resolution transmission electron microscopy (HRTEM), also known as phase contrast. When utilizing a Field emission source and a specimen of uniform thickness, the images are formed due to differences in phase of electron waves, which is caused by specimen interaction. Image formation is given by the complex modulus of the incoming electron beams. As such, the image is not only dependent on the number of electrons hitting the screen, making direct interpretation of phase contrast images more complex. However this effect can be used to an advantage, as it can be manipulated to provide more information about the sample, such as in complex phase retrieval techniques.

### *Diffraction*

As previously stated, by adjusting the magnetic lenses such that the back focal plane of the lens rather than the imaging plane is placed on the imaging apparatus a diffraction pattern can be generated. For thin crystalline samples, this produces an image that consists of a pattern of dots in the case of a single crystal, or a series of rings in the case of a polycrystalline or amorphous solid material. For the single crystal case the diffraction pattern is dependent upon the orientation of the specimen and the structure of the sample illuminated by the electron beam. This image provides the investigator with information about the space group symmetries in the crystal and the crystal's orientation to the beam path. This is typically done without utilising any information but the position at which the diffraction spots appear and the observed image symmetries.

Diffraction patterns can have a large dynamic range, and for crystalline samples, may have intensities greater than those recordable by CCD. As such, TEMs may still be equipped with film cartridges for the purpose of obtaining these images, as the film is a single use detector.

Analysis of diffraction patterns beyond point-position can be complex, as the image is sensitive to a number of factors such as specimen thickness and orientation, objective lens defocus, spherical and chromatic aberration. Although quantitative interpretation of the contrast shown in lattice images is possible, it is inherently complicated and can require extensive computer simulation and analysis, such as electron multislice analysis.

More complex behaviour in the diffraction plane is also possible, with phenomena such as Kikuchi lines arising from multiple diffraction within the crystalline lattice. In convergent beam electron diffraction (CBED) where a non-parallel, i.e. converging, electron wavefront is produced by concentrating the electron beam into a fine probe at the sample surface, the interaction of the convergent beam can provide information beyond structural data such as sample thickness.

### *Three-dimensional Imaging*

As TEM specimen holders typically allow for the rotation of a sample by a desired angle, multiple views of the same specimen can be obtained by rotating the angle of the sample along an axis perpendicular to the beam. By taking multiple images of a single TEM sample at differing angles, typically in 1° increments, a set of images known as a "tilt series" can be collected. This methodology was proposed in the 1970s by Walter Hoppe. Under purely absorption contrast conditions, this set of images can be used to construct a three-dimensional representation of the sample.

The reconstruction is accomplished by a two-step process, first images are aligned to account for errors in the positioning of a sample; such errors can occur due to vibration or mechanical drift. Alignment methods use image registration algorithms, such as autocorrelation methods to correct these errors. Secondly, using a technique known as filtered back projection, the aligned image slices can be transformed from a set of two-dimensional images, $I_j(x, y)$, to a single three-dimensional image, $I'_j(x, y, z)$. This three-dimensional image is of particular interest when morphological information is required, further study can be undertaken using computer algorithms, such as isosurfaces and data slicing to analyse the data.

As TEM samples cannot typically be viewed at a full 180° rotation, the observed images typically suffer from a "missing wedge" of data, which when using Fourier-based back projection methods decreases the range of resolvable frequencies in the three-dimensional reconstruction. Mechanical refinements, such as multi-axis tilting (two tilt series of the same specimen made at orthogonal directions) and conical tomography (where the specimen is first tilted to a given fixed angle and then imaged at equal angular rotational increments through one complete rotation in the plane of the specimen grid) can be used to limit the impact of the missing data on the observed specimen morphology. In addition, numerical techniques exist which can improve the collected data.

All the above-mentioned methods involve recording tilt series of a given specimen field. This inevitably results in the summation of a high dose of reactive electrons through the sample and the accompanying destruction of fine detail during recording. The technique of low-dose (minimal-dose) imaging is therefore regularly applied to mitigate this effect. Low-dose imaging is performed by deflecting illumination and imaging regions simultaneously away from the optical axis to image an adjacent region to the area to be recorded (the high-dose region). This area is maintained centred during tilting and refocused before recording. During recording the deflections are removed so that the area of interest is exposed to the electron beam only for the duration required for imaging. An improvement of this technique (for objects resting on a sloping substrate film) is to have two symmetrical off-axis regions for focusing followed by setting focus to the average of the two high-dose focus values before recording the low-dose area of interest.

Non-tomographic variants on this method, referred to as single particle analysis, use images of multiple (hopefully) identical objects at different orientations to produce the image data required for three-dimensional reconstruction. If the objects do not have significant preferred orientations, this method does not suffer from the missing data wedge (or cone) which accompany tomographic methods nor does it incur excessive radiation dosage, however it assumes that the different objects imaged can be treated as if the 3D data generated from them arose from a single stable object.

### *Sample Preparation*

Sample preparation in TEM can be a complex procedure. TEM specimens are required to be at most hundreds of nanometers thick, as unlike neutron or X-Ray radiation the electron beam interacts readily with the sample, an effect that increases roughly with atomic number squared ($z^2$). High quality samples will have a thickness that is comparable to the mean free path of the electrons that travel through the samples, which may be only a few tens of nanometers. Preparation of TEM specimens is specific to the material under analysis and the desired information to obtain from the specimen. As such, many generic techniques have been used for the preparation of the required thin sections.

Materials that have dimensions small enough to be electron transparent, such as powders or nanotubes, can be quickly prepared by the deposition of a dilute sample containing the specimen onto support grids or films. In the biological sciences in order to withstand the instrument vacuum and facilitate handling, biological specimens can be fixated using either a negative staining material such as uranyl acetate or by plastic embedding. Alternately samples may be held at liquid nitrogen temperatures after embedding in vitreous ice. In material science and metallurgy the specimens tend to be naturally resistant to vacuum, but still must be prepared as a thin foil, or etched so some portion of the specimen is thin enough for the beam to penetrate. Constraints on the thickness of the material may be limited by the scattering cross-section of the atoms from which the material is comprised.

### *Tissue Sectioning*

By passing samples over a glass or diamond edge, small, thin sections can be readily obtained using a semi-automated method. This method is used to obtain thin, minimally deformed samples that allow for the observation of tissue samples. Additionally inorganic samples have been studied, such as aluminium, although this usage is limited owing to the heavy damage induced in the less soft samples. To prevent charge build-up at the sample surface, tissue samples need to be coated with a thin layer of conducting material, such as carbon, where the coating thickness is several nanometers. This may be achieved via an electric arc deposition process using a sputter coating device.

### *Sample Staining*

Details in light microscope samples can be enhanced by stains that absorb light; similarly TEM samples of biological tissues can utilize high atomic number stains to enhance contrast. The stain absorbs electrons or scatters part of the electron beam which otherwise is projected onto the imaging system. Compounds of heavy metals such as osmium, lead, uranium or gold (in immunogold labelling) may be used prior to TEM observation to selectively deposit electron dense atoms in or on the sample in desired cellular or protein regions, requiring an understanding of how heavy metals bind to biological tissues.

### *Mechanical Milling*

Mechanical polishing may be used to prepare samples. Polishing needs to be done to a high quality, to ensure constant sample thickness across the region of interest. A diamond, or cubic boron nitride polishing compound may be used in the final stages of polishing to remove any scratches that may cause contrast fluctuations due to varying sample thickness. Even after careful mechanical milling, additional fine methods such as ion etching may be required to perform final stage thinning.

### *Chemical Etching*

Certain samples may be prepared by chemical etching, particularly metallic specimens. These samples are thinned using a chemical etchant, such as an acid, to prepare the sample for TEM observation.

Devices to control the thinning process may allow the operator to control either the voltage or current passing through the specimen, and may include systems to detect when the sample has been thinned to a sufficient level of optical transparency.

### Ion Etching

Ion etching is a sputtering process that can remove very fine quantities of material. This is used to perform a finishing polish of specimens polished by other means. Ion etching uses an inert gas passed through an electric field to generate a plasma stream that is directed to the sample surface. Acceleration energies for gases such as argon are typically a few kilovolts. The sample may be rotated to promote even polishing of the sample surface. The sputtering rate of such methods is on the order of tens of micrometres per hour, limiting the method to only extremely fine polishing. More recently focused ion beam methods have been used to prepare samples. FIB is a relatively new technique to prepare thin samples for TEM examination from larger specimens. Because FIB can be used to micro-machine samples very precisely, it is possible to mill very thin membranes from a specific area of interest in a sample, such as a semiconductor or metal. Unlike inert gas ion sputtering, FIB makes use of significantly more energetic gallium ions and may alter the composition or structure of the material through gallium implantation.

### Replication

Samples may also be replicated using cellulose acetate film, the film subsequently coated with a heavy metal, the original film melted away, and the replica imaged on the TEM. This technique is used for both materials and biological samples.

### Modifications

The capabilities of the TEM can be further extended by additional stages and detectors, sometimes incorporated on the same microscope. An *electron cryomicroscope* (CryoTEM) is a TEM with a specimen holder capable of maintaining the specimen at liquid nitrogen or liquid helium temperatures. This allows imaging specimens prepared in vitreous ice, the preferred preparation technique for imaging individual molecules or macromolecular assemblies. A TEM can be modified into a scanning transmission electron microscope (STEM) by the addition of a system that rasters the beam across the sample to form the image, combined with suitable detectors. Scanning coils are used to deflect the beam, such as by an electrostatic shift of the beam, where the beam is then collected using a current detector such as a

Faraday cup, which acts as a direct electron counter. By correlating the electron count to the position of the scanning beam (known as the "probe"), the transmitted component of the beam may be measured. The non-transmitted components may be obtained either by beam tilting or by the use of annular dark field detectors.

In-situ experiments may also be conducted with experiments such as in-situ reactions or material deformation testing. Modern research TEMs may include aberration correctors, to reduce the amount of distortion in the image. Incident beam Monochromators may also be used which reduce the energy spread of the incident electron beam to less than 0.15 eV. Major TEM makers include JEOL, Hitachi High-technologies, FEI Company (from merging with Philips Electron Optics), Carl Zeiss and NION.

## Low-voltage Electron Microscope

The low-voltage electron microscope (LVEM) is a combination of SEM, TEM and STEM in one instrument, which operated at relatively low electron accelerating voltage of 5 kV. Low voltage increases image contrast which is especially important for biological specimens. This increase in contrast significantly reduces, or even eliminates the need to stain. Sectioned samples generally need to be thinner than they would be for conventional TEM (20–65 nm). Resolutions of a few nm are possible in TEM, SEM and STEM modes.

### *Cryo-microscopy*

This technique allows TEM's to be used to see molecular structure of proteins and large molecules. Cryoelectron microscopy involves viewing unaltered macromolecular assemblies by vitrifying them, placing them on a grid and obtaining images by detecting electrons that transmit through the specimen.

### *Limitations*

There are a number of drawbacks to the TEM technique. Many materials require extensive sample preparation to produce a sample thin enough to be electron transparent, which makes TEM analysis a relatively time consuming process with a low throughput of samples. The structure of the sample may also be changed during the preparation process. Also the field of view is relatively small, raising the possibility that the region analyzed may not be characteristic of the whole sample. There is potential that the sample may be damaged by the electron beam, particularly in the case of biological materials.

### Resolution Limits

The limit of resolution obtainable in a TEM may be described in several ways, and is typically referred to as the information limit of the microscope. One commonly used value is a cut-off value of the contrast transfer function, a function that is usually quoted in the frequency domain to define the reproduction of spatial frequencies of objects in the object plane by the microscope optics. A cut-off frequency, $q_{max}$, for the transfer function may be approximated with the following equation, where $C_s$ is the spherical aberration coefficient and $\lambda$ is the electron wavelength:

$$q_{\max} = \frac{1}{0.67(C_s\lambda^3)^{1/4}}.$$

For a 200 kV microscope, with partly corrected spherical aberrations ("to the third order") and a $C_s$ value of 1 μm, a theoretical cut-off value might be $1/q_{max} = 42$ pm. The same microscope without a corrector would have $C_s = 0.5$ mm and thus a 200-pm cut-off. The spherical aberrations are suppressed to the third or fifth order in the "aberration-corrected" microscopes. Their resolution is however limited by electron source geometry and brightness and chromatic aberrations in the objective lens system. The frequency domain representation of the contrast transfer function may often have an oscillatory nature, which can be tuned by adjusting the focal value of the objective lens. This oscillatory nature implies that some spatial frequencies are faithfully imaged by the microscope, whilst others are suppressed. By combining multiple images with different spatial frequencies, the use of techniques such as focal series reconstruction can be used to improve the resolution of the TEM in a limited manner. The contrast transfer function can, to some extent, be experimentally approximated through techniques such as Fourier transforming images of amorphous material, such as amorphous carbon.

More recently, advances in aberration corrector design have been able to reduce spherical aberrations and to achieve resolution below 0.5 Ångströms (50 pm) at magnifications above 50 million times. Improved resolution allows for the imaging of lighter atoms that scatter electrons less efficiently, such as lithium atoms in lithium battery materials. The ability to determine the position of atoms within materials has made the HRTEM an indispensable tool for nanotechnology research and development in many fields, including heterogeneous catalysis and the development of semiconductor devices for electronics and photonics.

# 5

# Crystal Optics

Crystal optics is the branch of optics that describes the behaviour of light in *anisotropic media*, that is, media (such as crystals) in which light behaves differently depending on which direction the light is propagating. The index of refraction depends on both composition and crystal structure and can be calculated using the Gladstone–Dale relation. Crystals are often naturally anisotropic, and in some media (such as liquid crystals) it is possible to induce anisotropy by applying an external electric field.

## Isotropic Media

Typical transparent media such as glasses are *isotropic*, which means that light behaves the same way no matter which direction it is travelling in the medium. In terms of Maxwell's equations in a dielectric, this gives a relationship between the electric displacement field D and the electric field E:

$$D = \varepsilon_0 E + P$$

where $\varepsilon_0$ is the permittivity of free space and P is the electric polarization (the vector field corresponding to electric dipole moments present in the medium). Physically, the polarization field can be regarded as the response of the medium to the electric field of the light.

### *Electric Susceptibility*

In an isotropic and linear medium, this polarisation field P is proportional to and parallel to the electric field E:

$$P = \chi \varepsilon_0 E$$

where $\chi$ is the *electric susceptibility* of the medium. The relation between D and E is thus:

$$D = \varepsilon_0 E + \chi\varepsilon_0 E = \varepsilon_0(1+\chi)E = \varepsilon E$$

where

$$\varepsilon = \varepsilon_0(1+\chi)$$

is the dielectric constant of the medium. The value 1+$\chi$ is called the *relative permittivity* of the medium, and is related to the refractive index $n$, for non-magnetic media, by

$$n = \sqrt{1+\chi}$$

### *Anisotropic Media*

In an anisotropic medium, such as a crystal, the polarisation field P is not necessarily aligned with the electric field of the light E. In a physical picture, this can be thought of as the dipoles induced in the medium by the electric field having certain preferred directions, related to the physical structure of the crystal. This can be written as:

$$P = \varepsilon_0 \chi E.$$

Here $\chi$ is not a number as before but a tensor of rank 2, the *electric susceptibility tensor*. In terms of components in 3 dimensions:

$$\begin{pmatrix} P_x \\ P_y \\ P_z \end{pmatrix} = \varepsilon_0 \begin{pmatrix} \chi_{xx} & \chi_{xy} & \chi_{xz} \\ \chi_{yx} & \chi_{yy} & \chi_{yz} \\ \chi_{zx} & \chi_{zy} & \chi_{zz} \end{pmatrix} \begin{pmatrix} E_x \\ E_y \\ E_z \end{pmatrix}$$

or using the summation convention:

$$P_i = \varepsilon_0 \sum_{j\in\{x,y,z\}} \chi_{ij} E_j \quad .$$

Since $\chi$ is a tensor, P is not necessarily colinear with E.

In nonmagnetic and transparent materials, $\chi_{ij} = \chi_{ji}$, i.e. the $\chi$ tensor is real and symmetric. In accordance with the spectral theorem, it is thus possible to diagonalise the tensor by choosing the appropriate set of coordinate axes, zeroing all components of the tensor except $\chi_{xx}$, $\chi_{yy}$ and $\chi_{zz}$. This gives the set of relations:

$$P_x = \varepsilon_0 \chi_{xx} E_x$$

$$P_y = \varepsilon_0 \chi_{yy} E_y$$

$$P_z = \varepsilon_0 \chi_{zz} E_z$$

The directions x, y and z are in this case known as the *principal axes* of the medium. Note that these axes will be orthogonal if all entries in the $\chi$ tensor are real, corresponding to a case in which the refractive index is real in all directions.

It follows that D and E are also related by a tensor:

$$D = \varepsilon_0 E + P = \varepsilon_0 E + \varepsilon_0 \chi E = \varepsilon_0 (I + \chi) E = \varepsilon_0 \varepsilon E.$$

Here $\varepsilon$ is known as the *relative permittivity tensor* or *dielectric tensor*. Consequently, the refractive index of the medium must also be a tensor. Consider a light wave propagating along the z principal axis polarised such the electric field of the wave is parallel to the x-axis. The wave experiences a susceptibility $\chi_{xx}$ and a permittivity $\varepsilon_{xx}$. The refractive index is thus:

$$n_{xx} = (1 + \chi_{xx})^{1/2} = (\varepsilon_{xx})^{1/2}.$$

For a wave polarised in the y direction:

$$n_{yy} = (1 + \chi_{yy})^{1/2} = (\varepsilon_{yy})^{1/2}.$$

Thus these waves will see two different refractive indices and travel at different speeds. This phenomenon is known as *birefringence* and occurs in some common crystals such as calcite and quartz.

If $\chi_{xx} = \chi_{yy} \neq \chi_{zz}$, the crystal is known as uniaxial. If $\chi_{xx} \neq \chi_{yy}$ and $\chi_{xx} \neq \chi_{zz}$ the crystal is called biaxial. A uniaxial crystal exhibits two refractive indices, an "ordinary" index ($n_o$) for light polarised in the x or y directions, and an "extraordinary" index ($n_e$) for polarisation in the z direction. A uniaxial crystal is "positive" if $n_e > n_o$ and "negative" if $n_e < n_o$. Light polarised at some angle to the axes will experience a different phase velocity for different polarization components, and cannot be described by a single index of refraction. This is often depicted as an index ellipsoid.

### Other Effects

Certain nonlinear optical phenomena such as the electro-optic effect cause a variation of a medium's permittivity tensor when an external electric field is applied, proportional (to lowest order) to the strength of the field. This causes a rotation of the principal axes of the medium and alters the behaviour of light travelling through it; the effect can be used to produce light modulators. In response to a magnetic field, some materials can have a dielectric tensor that is complex-Hermitian; this is called a gyro-magnetic or magneto-optic effect. In this case, the principal axes are complex-valued vectors, corresponding to elliptically polarized light, and time-reversal symmetry can be broken. This can be used to design optical isolators, for example.

A dielectric tensor that is not Hermitian gives rise to complex eigenvalues, which corresponds to a material with gain or absorption at a particular frequency.

## Polarized Light Microscopy

Polarized light microscopy can mean any of a number of optical microscopy techniques involving polarized light. Simple techniques include illumination of the sample with polarized light. Directly transmitted light can, optionally, be blocked with a polariser orientated at 90 degrees to the illumination. More complex microscopy techniques which take advantage of polarized light include differential interference contrast microscopy and interference reflection microscopy.

These illumination techniques are most commonly used on birefringent samples where the polarized light interacts strongly with the sample and so generating contrast with the background. Polarized light microscopy is used extensively in optical mineralogy. Polarized light microscopy is capable of providing information on absorption colour and optical path boundaries between minerals of differing refractive indices, in a manner similar to brightfield illumination, but the technique can also distinguish between isotropic and anisotropic substances. Furthermore, the contrast-enhancing technique exploits the optical properties specific to anisotropy and reveals detailed information concerning the structure and composition of materials that are invaluable for identification and diagnostic purposes.

### *Basic Properties of Polarized Light*

The wave model of light describes light waves vibrating at right angles to the direction of propagation with all vibration directions being equally probable. This is referred to as "common" or "non-polarized" white light. In polarized light there is only one vibration direction. The human eye-brain system has no sensitivity to the vibration directions of light, and polarized light can only be detected by an intensity or colour effect, for example, by reduced glare when wearing polarized sun glasses.

Polarized light is most commonly produced by absorption of light having a set of specific vibration directions in a dichroic medium. Certain natural minerals, such as tourmaline, possess this property, but synthetic films invented by Dr. Edwin H. Land in 1932 soon overtook all other materials as the medium of choice for production of polarized light. Tiny crystallites of iodoquinine sulphate, oriented in the same direction, are embedded in a transparent polymeric film to prevent migration and reorientation of the crystals. Land developed sheets containing polarizing films that were marketed under the trade name of Polaroid, which has become the accepted generic term for these sheets. Any device capable of selecting polarized light from natural (unpolarized) white light is now referred to as a polar or polarizer, a name first introduced in 1948 by A. F. Hallimond. Today, polarizers are widely used in liquid crystal displays (LCDs), sunglasses, photography, microscopy, and for a myriad of scientific and medical purposes.

There are two polarizing filters in a polarizing microscope - termed the polarizer and analyzer. The polarizer is positioned beneath the specimen stage usually with its vibration azimuth fixed in the left-to-right, or East-West direction, although most of these elements can be rotated through 360 degrees. The analyzer, usually aligned with a vibration direction oriented North-South, but again rotatable on some microscopes, is placed above the objectives and can be moved in and out of the light path as required. When both the analyzer and polarizer are inserted into the optical path, their vibration azimuths are positioned at right angles to each other. In this configuration, the polarizer and analyzer are said to be crossed, with no light passing through the system and a dark viewfield present in the eyepieces.

For incident light polarized microscopy, the polarizer is positioned in the vertical illuminator and the analyzer is placed above the half mirror. Most rotatable polarizers are graduated to indicate the rotation angle of the transmission azimuth, while analyzers are usually fixed into position (although advanced models can be rotated either 90 or 360 degrees). The polarizer and analyzer are the essential components of the polarizing microscope, but other desirable features include:

- Specialized Stage - A 360-degree circular rotating specimen stage to facilitate orientation studies with centration of the objectives and stage with the microscope optical axis to make the centre of rotation coincide with the centre of the field of view. Many stages designed for polarized light microscopy also contain a vernier scale so that rotation angle can be measured to an accuracy of 0.1 degree. For advanced studies of conoscopic images, a universal stage having multiple axes of rotation can also be employed to enable observation of the specimen from any direction.
- Strain Free Objectives - Stress introduced into the glass of an objective during assembly can produce spurious optical effects under polarized light, a factor that could compromise

performance. Objectives designed for polarized light observation are distinguished from ordinary objectives with the inscription P, PO, or Pol on the barrel. The performance of an objective is limited by several factors, including the anti-reflection coatings used on lens surfaces, and the refractive properties due to angle of incident light on the front lens. In addition, lens strain can be introduced at the cement junction between elements in a lens group or from a single or group of lenses that has been mounted too tightly in the frame.

- Centerable Revolving Nosepiece - Because the objective optical axis position varies from one assembly to another, many polarized light microscopes are equipped with a specialized nosepiece that contains a centering mechanism for individual objectives. This enables each objective to be centred with respect to the stage and microscope optical axis so that specimen features remain in the centre of the viewfield when the stage is rotated through 360 degrees.
- Strain Free Condenser - Condensers designed for polarized light microscopy have several features in common, including the use of strain free lenses. Some condensers are equipped with a receptacle for the polarizer or have the polarizing element mounted directly into the condenser, beneath the aperture diaphragm. Many polarized light condensers have a top lens that can be removed (a swing-lens condenser) from the light path to generate nearly parallel illumination wavefronts for low magnification and birefringence observations.
- Eyepieces - Polarized light microscope eyepieces are fitted with a cross wire reticle (or graticule) to mark the centre of the field of view. Often, the cross wire reticle is substituted for a photomicrography reticle that assists in focusing the specimen and composing images with a set of frames bounding the area of the viewfield to be captured either digitally or onto film. Orientation of the eyepiece with respect to the polarizer and analyzer is guaranteed by a point pin that slides into the observation tube sleeve.
- Bertrand Lens - A specialized lens mounted in an intermediate tube or within the observation tubes, a Bertrand lens projects an interference pattern formed at the objective rear focal plane into focus at the microscope image plane. The lens is designed to enable easy examination of the objective rear focal plane,

to allow accurate adjustment of the illuminating aperture diaphragm and to view interference figures, similar to the ones presented. Note that in figures, the interference patterns represent those observed with a uniaxial crystal in polarized light, while the pattern is typical of a uniaxial crystal with a first order retardation plate inserted into the optical pathway.

- Compensator and Retardation Plates - Many polarized light microscopes contain a slot to allow the insertion of compensators and/or retardation plates between the crossed polarizers, which are used to enhance optical path differences in the specimen. In most modern microscope designs, this slot is placed either in the microscope nosepiece or an intermediate tube positioned between the body and eyepiece tubes. Compensation plates inserted into the slot are then situated between the specimen and the analyzer.

Polarized light microscopy can be used both with reflected (incident or epi) and transmitted light. Reflected light is useful for the study of opaque materials such as ceramics, mineral oxides and sulfides, metals, alloys, composites, and silicon wafers. Reflected light techniques require a dedicated set of objectives that have not been corrected for viewing through the cover glass, and those for polarizing work should also be strain free.

### *The Michel-Levy Chart*

As polarised light passes through a birefringent sample, the phase difference between the fast and slow directions varies with the thickness, and wavelength of light used. The optical path difference (o.p.d.) is defined as $o.p.d. = \Delta n{\cdot}t$, where t is the thickness of the sample.

This then leads to a phase difference between the light passing in the two vibration directions of $\delta = 2\pi(\Delta n{\cdot}t / \lambda)$. For example, if the optical path difference is $\lambda/2$, then the phase difference will be $\pi$, and so the polarisation will be perpendicular to the original, resulting in all of the light passing through the analyser for crossed polars. If the optical path difference is $n{\cdot}\lambda$, then the phase difference will be $2n{\cdot}\pi$, and so the polarisation will be parallel to the original. This means that no light will be able to pass though the analyser which it is now perpendicular to.

The Michel-Levy Chart arises when polarised white light is passed through a birefringent sample. If the sample is of uniform thickness, then only one specific wavelength will meet the above condition described above, and be perpendicular to the direction of the analyser. This means that instead of polychromatic light being viewed at the analyser, one specific wavelength will have been removed. This information can be used in a number of ways:

- If the birefringence is known, then the thickness, t, of the sample can be determined
- If the thickness is known, then the birefringence of the sample can be determined

As the order of the optical path difference increases, then it is more likely that more wavelengths of light will be removed from the spectrum. This results in the appearance of the colour being "washed out", and it becomes more difficult to determine the properties of the sample. This, however, only occurs when the sample is relatively thick when compared to the wavelength of light.

## Conoscopy

Conoscopy "cone, spinning top, pine cone" and skopeo "examine, inspect, look to or into, consider" is an optical technique to make observations of a transparent specimen in a cone of converging rays of light. The various directions of light propagation are observable simultaneously .

A conoscope is an apparatus to carry out *conoscopic observations* and measurements, often realized by a microscope with a Bertrand lens for observation of the *directions image*. The earliest references on the use of *conoscopy* (i.e., observation in convergent light with a polarization microscope with a Bertrand lens) for evaluation of the optical properties of liquid crystalline phases (i.e., orientation of the optical axes) dates back to 1911 when it was used by Mauging to investigate the alignment of nematic and chiral-nematic phases.

A beam of convergent (or divergent) light is known to be a linear superposition of many plane waves over a cone of solid angles. The raytracing of figure  illustrates the basic concept of *conoscopy*: transformation of a directional distribution of rays of light in the front focal plane into a lateral distribution (*directions image*) appearing in the back focal plane (which is more or less curved). The incoming elementary parallel beams (illustrated by the colours blue, green and red) are converging in the back focal plane of the lens with the distance of their focal point from the optical axis being a (monotonous) function of the angle of beam inclination.

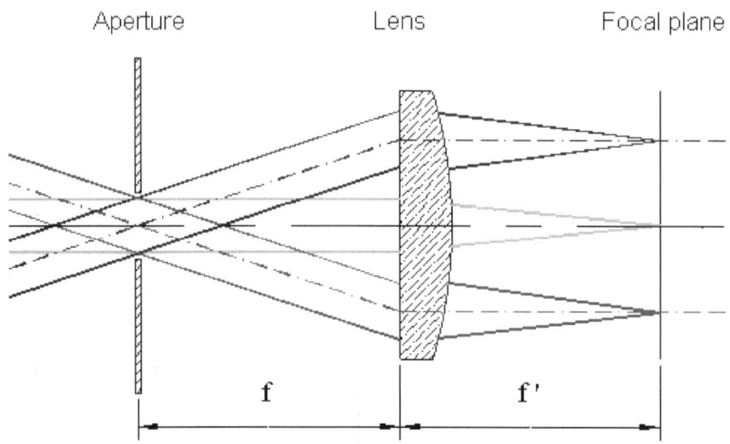


***Figure:*** *Imaging of bundles of elementary parallel rays to form a directions image in the back focal plane of a positive thin lens.*

This transformation can easily be deduced from two simples rules for the thin positive lens:

- the rays through the centre of the lens remain unchanged,
- the rays through the front focal point are transformed into parallel rays.

The object of measurement is usually located in the front focal plane of the lens. In order to select a specific area of interest on the object (i.e., definition of a measuring spot, or field of measurement) an aperture can be placed on top of the object. In this configuration only rays from the measuring spot (aperture) hit the lens.

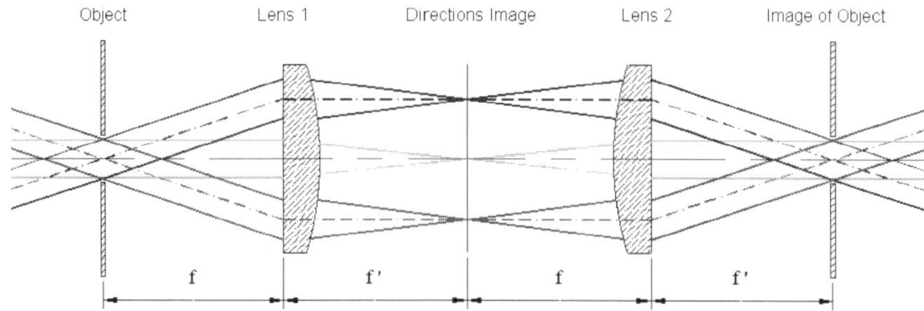


***Figure:*** *Formation of an image of the object (aperture) by addition of a second lens. The field of measurement is determined by the aperture located in the image of the object.*

The image of the aperture is projected to infinity while the image of the directional distribution of the light passing through the aperture

(i. e. directions image) is generated in the back focal plane of the lens. When it is not considered appropriate to place an aperture into the front focal plane of the lens, i.e., on the object, the selection of the measuring spot (field of measurement) can also be achieved by using a second lens. An image of the object (located in the front focal plane of the first lens) is generated in the back focal plane of the second lens. The magnification, M, of this imaging is given by the ratio of the focal lengths of the lenses $L_1$ and $L_2$, $M = f_2 / f_1$.

A third lens transforms the rays passing through the aperture (located in the plane of the image of the object) into a second directions image which may be analyzed by an image sensor (e.g., electronic camera).

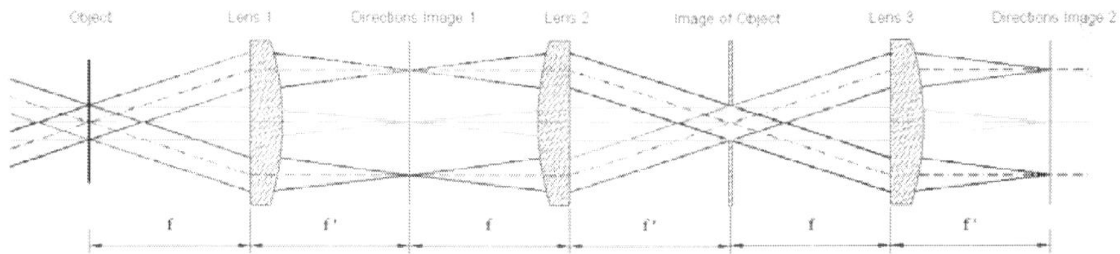


***Figure:*** *Schematic raytracing of a complete conoscope: formation of the directions image and imaging of the object.*

The functional sequence is as follows:

- the first lens forms the directions image (transformation of directions into locations),
- the second lens together with the first projects an image of the object,
- the aperture allows selection of the area of interest (measuring spot) on the object,
- the third lens together with the second images the directions image on a 2-dimensional optical sensor (e.g., electronic camera).

This simple arrangement is the basis for all conoscopic devices (conoscopes). It is not straight forward however to design and manufacture lens systems that combine the following features:

- maximum angle of light incidence as high as possible (e.g., 80°),
- diametre of measuring spot up to several millimetres,
- achromatic performance for all angles of inclination,
- minimum effect of polarization of incident light.

Design and manufacturing of this type of complex lens system requires assistance by numerical modelling and a sophisticated manufacturing process.

Modern advanced conoscopic devices are used for rapid measurement and evaluation of the electro-optical properties of LCD-screens (e.g., variation of luminance, contrast and chromaticity with viewing direction).

## Silicate Minerals

The silicate minerals make up the largest and most important class of rock-forming minerals, constituting approximately 90 percent of the crust of the Earth. They are classified based on the structure of their silicate group which contain different ratios of silicon and oxygen.

### *Abswurmbachite*

Abswurmbachite is a copper manganese silicate mineral ($(Cu,Mn^{2+})Mn^{3+}{}_6O_8SiO_4$). It was first described in 1991 and named after Irmgard Abs-Wurmbach (born 1938), a German mineralogist. It crystallizes in the tetragonal system. Its Mohs scale rating is 6.5 and a specific gravity of 4.96. It has a metallic luster and its colour is jet black, with light brown streaks.

### *Afwillite*

Afwillite is a calcium hydroxide nesosilicate mineral with formula $Ca_3(SiO_3OH)_2 \cdot 2H_2O$. It occurs as glassy, colourless to white prismatic monoclinic crystals. Its Mohs scale hardness is between 3 and 4. It occurs as an alteration mineral in contact metamorphism of limestone. It occurs in association with apophyllite, natrolite, thaumasite, merwinite, spurrite, gehlenite, ettringite, portlandite, hillebrandite, foshagite, brucite and calcite.

It was first described in 1925 for an occurrence in the Dutoitspan Mine, Kimberley, South Africa and was named for Alpheus Fuller Williams (1874–1953), a past official of the De Beers diamond company.

Afwillite is typically found in veins of spurrite and it belongs to the nesosilicate sub-class. It is monoclinic, its space group is P2 and its point group is 2.

### *Formation of Afwillite*

It is suggested that afwillite forms in fractured veins of the mineral spurrite. Jennite, afwillite, oyelite and calcite are all minerals that form in layers within spurrite veins. It appears that the afwillite, as well as the calcite, forms from precipitated fluids. The jennite is actually an alteration of the afwillite, but both formed from calcium silicates through hydration. Laboratory studies determined that afwillite forms at a temperature below 200 °C, usually around 100 °C. Afwillite and spurrite are formed through contact metamorphism of limestone. Contact metamorphism is caused by the interaction of rock with heat and/or fluids from a nearby crystallizing silicate magma.

## Structure and Properties

Afwillite has a complex monoclinic structure, and the silicon tetrahedra in the crystal structure are held together by hydrogen bonds. It has perfect cleavage parallel to its (101) and poor cleavage parallel to its (100) faces. It is biaxial and its 2V angle, the measurement from one optical axis to the other optical axis, is 50 – 56 degrees. When viewed under crossed polarizers in a petrographic microscope, it displays first-order orange colours, giving a maximum birefringence of 0.0167 (determined by using the Michel- Levy chart). Afwillite is optically positive. Additionally, it has a prismatic crystal habit. Under a microscope afwillite looks like wollastonite, which is in the same family as afwillite.

Awfillite is composed of double chains that consist of calcium and silicon polyhedral connected to each other by sharing corners and edges. This causes continuous sheets to form parallel to its miller index [-101] faces. The sheets are bonded together by hydrogen bonds and are all connected by Ca-Si-O bonds (Malik and Jeffery 1976). Each calcium atom is in 6-fold octahedral coordination with the oxygen, and the silicon is in 4-fold tetrahedral coordination around the oxygen. Around each silicon there is one OH group and there are three oxygens that neighbour them. The silicon tetrahedra are arranged so that they share an edge with calcium(1), and silicon(2) shares edges with the calcium(2) and calcium(3) polyhedral. The silicon tetrahedra are held together by the OH group and hydrogen bonding occurs between the hydrogen in the OH and the silicon tetrahedra. Hydrogen bonding is caused because the positive ion, hydrogen, is attracted to negatively charge ions which, in this case, are the silicon tetrahedra.

### *Occurrence in Concrete*

Afwillite is one of the calcium silicates that form when Portland cement sets to form concrete. The cement gets its strength from the hydration of its di- and tri- calcium silicates.

## Aluminosilicate

Aluminosilicate minerals are minerals composed of aluminium, silicon, and oxygen, plus countercations. They are a major component of kaolin and other clay minerals.

Andalusite, kyanite, and sillimanite are naturally occurring aluminosilicate minerals that have the composition $Al_2SiO_5$. The triple point of the three polymorphs is located at a temperature of 500 °C and a pressure of 0.4 GPa. These three minerals are commonly used as index minerals in metamorphic rocks.

Hydrated aluminosilicate minerals are referred to as zeolites and are porous structures that are naturally occurring materials.

The catalyst silica-alumina is an amorphous substance which is not an aluminosilicate compound.

### *Andalusite*

Andalusite is an aluminium nesosilicate mineral with the chemical formula $Al_2SiO_5$.

The variety chiastolite commonly contains dark inclusions of carbon or clay which form a checker-board pattern when shown in cross-section.

A clear variety first found in Andalusia, Spain can be cut into a gemstone. Faceted andalusite stones give a play of red, green, and yellow colours that resembles a muted form of iridescence, although the colours are actually the result of unusually strong pleochroism.

It is associated with mica schist which increases alkali content in ultimate product and so it has not been exploited economically so far.

### *Occurrence*

Andalusite is a common regional metamorphic mineral which forms under low pressure and low to high temperatures. The minerals kyanite and sillimanite are polymorphs of andalusite, each occurring under different temperature-pressure regimes and are therefore rarely found together in the same rock. Because of this the three minerals are a useful tool to help identify the pressure-temperature paths of the host rock in which they are found. An example rock includes hornfels.

It was first described and named after the type locality in the Ronda Massif, Málaga, Andalusia, Spain in 1789.

### *Bakerite*

Bakerite is the common name given to hydrated calcium borosilicate hydroxide, a borosilicate mineral (chemical formula $Ca_4B_4(BO_4)(SiO_4)_3(OH)_3{\cdot}(H_2O)$) that occurs in volcanic rocks in the Baker, California area.

It was first described in 1903 for an occurrence in the Corkscrew Canyon Mine of the Black Mountains, Furnace Creek District, Death Valley National Park, Inyo County, California, USA. It was named for Richard C. Baker, a director of the Pacific Coast Borax Company.

### *Braunite*

Braunite is a silicate mineral containing both di- and tri-valent manganese with the chemical formula: $Mn2{+}Mn^{3+}{}_6[O_8 \mid SiO_4]$. Common impurities include iron, calcium, boron, barium, titanium, aluminium, and magnesium.

Braunite forms grey/black tetragonal crystals and has a Mohs hardness of 6 - 6.5.

It was named after the Wilhelm von Braun (1790–1872) of Gotha, Thuringia, Germany.

A calcium iron bearing variant, named braunite II (formula: $Ca(Mn^{3+},Fe^{3+})_{14}SiO_{24}$), was discovered and described in 1967 from Kalahari, Cape Province, South Africa.

## Bultfonteinite

Bultfonteinite, originally dutoitspanite, is a pink to colourless mineral with chemical formula $Ca_2SiO_2(OH,F)_4$. It was discovered in 1903 or 1904 in the Bultfontein mine in South Africa, for which the mineral is named, and described in 1932.

### *Description*

Bultfonteinite is transparent and pale pink to colourless. The mineral occurs as radiating prismatic acicular crystals and radial spherules up to 2 cm (0.79 in).

### *Structure*

The crystal structure of bultfonteinite consists of strips of $[Ca_4Si_2O_4]3+$, that run along the 5.67 Å *c*-axis, held together by Ca–O–Ca, Ca–F–Ca, Ca–$H_2O$–Ca, and Ca–O–Si bonds. Silicon atoms occur in isolated tetrahedra and the calcium atoms have seven-fold coordination, derived from a triangular prism with a seventh atom present on one of the square faces.

### History

In either 1903 or 1904, a miner discovered the first specimen of bultfonteinite on the 480-foot level of the Bultfontein mine in Kimberley, South Africa. The mineral occurred in a several-hundred-foot-tall horse of kimberlite-enclosed dolerite and shale fragments. The specimen, mistakenly thought to be natrolite, was given to Alpheus F. Williams. Several years later, additional samples were found by C. E. Adams in the nearby Dutoitspan mine and given to the MacGregor Museum in Kimberley. Shortly before 1932, the mineral was found about 100 miles (160 km) to the southeast of Kimberley at the Jagersfontein Mine in Orange River Colony.

After John Parry and F. E. Wright described the mineral afwillite in 1925, Williams recognized that his samples of bultfonteinite were not natrolite, but were likely a new mineral species. Chemical analysis by John Parry and crystallographic and optical determination by Wright proved it to be a new mineral. The mineral was described by Parry, Williams, and Wright in 1932 and named *bultfonteinite.* Their original description does not explicitly state the origin of the name, but it is presumably named after the mine in which it was discovered. Earlier that year in his book *The Genesis of the Diamond,* Williams had called the mineral dutoitspanite, a name which was "apparently discarded". When the International Mineralogical Association was founded, bultfonteinite was grandfathered as a valid mineral species.

The type material is held in England at Cambridge University and the Natural History Museum in London.

### Occurrence

Bultfonteinite has been found in Australia, Botswana, Canada, Israel, Japan, Jordan, Russia, South Africa, and the United States. The mineral was first located outside South Africa in the US State of California in 1955. Bultfonteinite has been found in association with afwillite, apophyllite, calcite, natrolite, oyelite, scawtite, and xonotlite.

At the type locality, the mineral occurred in a large structure of dolerite and shale fragments in a kimberlite pipe. In Crestmore, California, bultfonteinite formed in the contact zone of thermally metamorphosed limestone.

### Cerite

Cerite is a complex silicate mineral group containing cerium, formula $(Ce,La,Ca)_9(Mg,Fe+3)(SiO_4)_6(SiO_3OH)(OH)_3$. The cerium and lanthanum content varies with the Ce rich species (cerite-(Ce)) and the La rich species (cerite-(La)). Analysis of a sample from the Mountain Pass carbonatite gave 35.05% $Ce_2O_3$ and 30.04% $La_2O_3$.

Cerite was first described in 1803 for an occurrence in Bastnäs in Västmanland, Sweden. The lanthanum rich species, cerite-(La) was first described for an occurrence in the Khibina massif, Kola Peninsula, Russia in 2002.

## Chiastolite

The mineral chiastolite is a variety of andalusite with the chemical composition Al2SiO5. It is noted for distinctive cross-shaped black inclusions of graphite. In areas around Georgetown, California, metamorphosed sediments contained andalusite and chiastolite in a graphite rich metasediment. The chiastolite crystals have been pseudomorphically altered by a mixture of muscovite, paragonite and margarite. The calcium rich margarite tends to form along the graphite rich crosses or bands within the chiastolite. Mineralogically the occurrence is important because all three white mica phases are present in an equilibrium assemblage.

## *Chloritoid*

Chloritoid is a silicate mineral of metamorphic origin. It is an iron magnesium manganese alumino-silicate hydroxide with formula: $(Fe,Mg,Mn)_2Al_4Si_2O_{10}(OH)_4$. It occurs as greenish grey to black platy micaceous crystals and foliated masses. Its Mohs hardness is 6.5, unusually high for a platy mineral, and it has a specific gravity of 3.52 to 3.57. It typically occurs in phyllites, schists and marbles.

Both monoclinic and triclinic polytypes exist and both are pseudohexagonal.

It was first described in 1837 from localities in the Ural Mountains region of Russia. It was named for its similarity to the chlorite group of minerals.

## *Clinohedrite*

Clinohedrite is a rare silicate mineral. Its chemical composition is a hydrous calcium-zinc silicate; $CaZn(SiO)_4 \cdot H_2O$. It crystallizes in the monoclinic system and typically occurs as veinlets and fracture coatings. It is commonly colourless, white to pale amethyst in colour. It has perfect cleavage and the crystalline habit has a brilliant luster. It has a Mohs hardness of 5.5 and a specific gravity of 3.28 - 3.33.

Under short wave ultraviolet light it fluoresces a rich orange colour. It is frequently associated with minerals such as hardystonite (fluoresces violet blue), esperite (fluoresces bright yellow), calcite (fluoresces orange-red), franklinite (non-fluorescent) and willemite (fluoresces green).

Clinohedrite was found primarily at the Franklin zinc mines in New Jersey, the type locality, but has also been reported from the Christmas mine, Gila County, Arizona, and the Western Quinling gold belt, Gansu Province, China.

It was first described in 1898 and was named for its crystal morphology from the Greek *klino* for incline, and *hedra* for face.

### Coffinite

Coffinite is a uranium-bearing silicate mineral with formula: $U(SiO_4)_{1-x}(OH)_{4x}$.

It occurs as black incrustations, dark to pale-brown in thin section. It has a grayish black streak. It has a brittle to conchoidal fracture. The hardness of coffinite is between 5 and 6.

It was first described in 1954 for an occurrence at the La Sal No. 2 Mine, Beaver Mesa, Mesa County, Colorado, USA, and named for American geologist Reuben Clare Coffin (1886–1972). It has widespread global occurrence in Colorado Plateau-type uranium ore deposits of uranium and vanadium. It replaces organic matter in sandstone and in hydrothermal vein type deposits. It occurs in association with uraninite, thorite, pyrite, marcasite, roscoelite, clay minerals and amorphous organic matter.

### Composition

Coffinite's chemical formula is $U(SiO_4)_{1-x}(OH)_{4x}$. X-ray powder patterns from samples of coffinite allowed geologists to classify it as a new mineral in 1955. A comparison to the x-ray powder pattern of zircon ($ZrSiO_4$) and thorite ($ThSiO_4$) was the basis for this classification. Preliminary chemical analysis indicated that the uranous silicate exhibited hydroxyl substitution. The results of Sherwood's preliminary chemical analysis were based on samples from three locations. Hydroxyl bonds and silicon-oxygen bonds also proved to exist after infrared absorption spectral analyses were performed. The hydroxyl substitution occurs as $(OH)_4^{4-}$ for $(SiO_4)^{2-}$. The hydroxyl constituent in coffinite later proved to be nonessential in the formation of a stable synthetic mineral. Recent electron microprobe analysis of the submicroscopic crystals uncovered an abundance of calcium, yttrium, phosphorus, and minimal lead substitutions along with traces of other rare earth elements.

### *Crystal Structure*

Coffinite is isostructural with the orthosilicates zircon ($ZrSiO_4$) and thorite ($ThSiO_4$). Stieff et al. analyzed coffinite using the x-ray powder diffraction technique and determined that it has a tetragonal structure. Occurring naturally with $U^{4+}$ cations, the $UO_8$ triangular dodecahedra coordinate with edge-sharing, alternating $SiO_4$ tetrahedra in chains along the c-axis. The central uranium site of coffinite is surrounded by eight $SiO_4$ tetrahedra. The lattice dimensions of naturally-occurring and synthetic coffinite are similar, with a naturally-occurring sample from Arrowhead Mine, Mesa County, Colorado having a=6.93kx, c=6.30kx, and a sample synthesized by Hoekstra and Fuchs having a=6.977kx and c=6.307kx.

### *Physical Properties*

Initial examination of coffinite by Stieff et al. described the mineral as black in colour with an adamantine luster, indistinguishable from uraninite ($UO_2$). Additionally, the discoverers reported that although no cleavage is seen in coffinite, it does exhibit subconchoidal fracturing and is very fine grained. Initial samples showed a brittle texture and a hardness between 5 and 6, with a specific gravity of 5.1. Later samples from Woodrow Mine in New Mexico collected by Moench showed fibrous internal structure and exceptional crystallization. A polished thin section of coffinite has a brown colour and shows anisotropic transmission of light. Optical analysis yielded a refractive index of about 1.74.

### *Geological Occurrence*

Coffinite was first discovered in sedimentary uranium deposits in the Colorado Plateau region, but has also been discovered in sedimentary uranium deposits and hydrothermal veins in many other locations. Samples of coffinite from the Colorado Plateau were found with black fine-grained low-valence vanadium minerals, uraninite and finely dispersed black organic material. Other materials associated with later finds from the same region were clay and quartz. In vein deposits of the Copper King Mine in Colorado, coffinite was also found to occur with uraninite and pitchblende. Formation of coffinite requires a uranium source, and may happen in reducing conditions, as evidenced by the associated presence of low-valence vanadium minerals. Silica-rich solution provides such a reducing condition in cases where coffinite results as an alteration product of uraninite. Hansley and Fitzpatrick also noted that the brownish colour of their coffinite samples was caused by organic material, leading them to conclude that coffinite can also form in low temperature conditions if organic carbon is present. This finding is consistent with the coffinite samples of the Colorado Plateau, which included fossilized wood. In China, coffinite can be found in granite in addition to sandstone. Hansley and Fitzpatrick concluded that coarse-grained coffinite most likely forms in high temperature environments. Coffinite and uraninite precipitate inside brecciated and fractured regions of altered granite at pressures between 500 to 800 bars and temperatures at 126 to 178 °C.

### *Special Characteristics*

A large percentage of the Earth's uranium supply is contained in coffinite deposits, which is significant because of uranium's uses in nuclear energy and weaponry applications. Sedimentary deposits contain the most radioactive samples, as evidenced by the intensely radioactive coffinite found in the Colorado Plateau. Researchers at Harvard University, the United States Geological Survey (USGS), and several other institutions attempted unsuccessfully to synthesize coffinite in the mid-1950s after its initial discovery. In 1956, Hoekstra and Fuchs managed to create stable samples of synthetic coffinite. All of this research was conducted for the United States Atomic Energy Commission.

## Gadolinite

Gadolinite, sometimes also known as Ytterbite, is a silicate mineral which consists principally of the silicates of cerium, lanthanum, neodymium, yttrium, beryllium, and iron with the formula (Ce,La,Nd,Y)2FeBe2Si2O10. It is called gadolinite-(Ce) or gadolinite-(Y) depending on the prominence of the variable element composition (namely, Y if it has more yttrium, and Ce if it has more cerium). It may contain 35.48% yttria sub-group rare earths, 2.17% ceria earths, up to 11.6% BeO and traces of thorium. It is found in Sweden, Norway, and the USA (Texas and Colorado).

### *Characteristics*

Gadolinite is fairly rare and typically occurs as well-formed crystals. It is nearly black in colour and has a vitreous luster. Its hardness is between 6.5 and 7, and its specific gravity is between 4.0 and 4.7. It fractures in a conchoidal pattern. The mineral's streak is grayish-green. It is also pyrognomic, which means that it becomes incandescent at a relatively low temperature.

### Name and Discovery

Gadolinite was named in 1800 for Johan Gadolin, the Finnish mineralogist-chemist who first isolated an oxide of the rare earth element yttrium from the mineral in 1792. The rare earth gadolinium was also named for him. However, gadolinite does not contain more than trace amounts of gadolinium. When Gadolin analyzed this mineral, he missed an opportunity to discover a second element: what he thought was aluminium (alumina) was in fact an element that would not be officially discovered until 1798: beryllium (beryllia).

### Uses

Gadolinite and euxenite are quite abundant and are future sources of yttrium sub group rare earths. At present, these elements are recovered from monazite concentrates (after recovery of ceria sub-group metals).

## Forsterite

Forsterite ($Mg_2SiO_4$) is the magnesium rich end-member of the olivine solid solution series. Forsterite crystallizes in the orthorhombic system (space group *Pbnm*) with cell parametres *a* 4.75 Å (0.475 nm), *b* 10.20 Å (1.020 nm) and *c* 5.98 Å (0.598 nm).

Forsterite is associated with igneous and metamorphic rocks and has also been found in meteorites. In 2005 it was also found in cometary dust returned by the Stardust probe. In 2011 it was observed as tiny crystals in the dusty clouds of gas around a forming star.

Two polymorphs of forsterite are known: wadsleyite (also orthorhombic) and ringwoodite (isometric). Both are mainly known from meteorites.

Peridot is the gemstone variety of forsterite olivine.

Forsterite reacts with quartz to form the orthopyroxene mineral enstatite in the following reaction:

$$Mg_2SiO_4 + SiO_2 \rightarrow 2\ MgSiO_3.$$

### Composition

Pure forsterite is composed of magnesium, oxygen and silicon. The chemical formula is $Mg_2SiO_4$. Forsterite, fayalite ($Fe_2SiO_4$) and tephroite ($Mn_2SiO_4$) are the end-members of the olivine solid solution series; other elements such as Mn, Ni, and Ca substitute for Fe and Mg in olivine, but only in minor proportions in natural occurrences. Other minerals such as monticellite ($CaMgSiO_4$), an uncommon calcium-rich mineral, share the olivine structure, but solid solution between olivine and these other minerals is limited. Monticellite is found in contact metamorphosed dolomites.

### Geologic Occurrence

Forsterite-rich olivine is the most abundant mineral in the mantle above a depth of about 400 km; pyroxenes are also important minerals in this upper part of the mantle. Although pure forsterite does not occur in igneous rocks, dunite often contains olivine with forsterite contents at least as Mg-rich as $Fo_{92}$ (92% forsterite – 8% fayalite); common peridotite contains olivine typically at least as Mg-rich as $Fo_{88}$. Forsterite-rich olivine is a common crystallization product of mantle-derived magma. Olivine in mafic and ultramafic rocks typically is rich in the forsterite end-member.

Forsterite also occurs in dolomitic marble which results from the metamorphism of high magnesium limestones and dolostones. Nearly pure forsterite occurs in some metamorphosed serpentinites. Fayalite-rich olivine is much less common. Nearly pure fayalite is a minor constituent in some granite-like rocks, and it is a major constituent of some metamorphic banded iron formations.

### Structure and Formation

Forsterite is mainly composed of the anion $SiO_4^{4-}$ and the cation $Mg^{2+}$ in a molar ratio 1:2. Silicon is the central atom in the $SiO_4^{4-}$ anion. Each oxygen atom is bonded to the silicon by a single covalent bond. The four oxygen atoms have a partial negative charge because of the covalent bond with silicon. Therefore, oxygen atoms need to stay far from each other in order to reduce the repulsive force between them. The best geometry to reduce the repulsion is a tetrahedral shape. The cations occupy two different octahedral sites which are M1 and M2 and form ionic bonds with the silicate anions. M1 and M2 are slightly different. M2 site is larger and more regular than M1 as shown in the figure. The packing in forsterite structure is dense. The space group of this structure is Pbnm and the point group is 2/m 2/m 2/m which is an orthorhombic crystal structure.

his structure of forsterite can form a complete solid solution by replacing the magnesium with iron. Iron can form two different cations which are $Fe^{2+}$ and $Fe^{3+}$. The iron(II) ion has the same charge as magnesium ion and it has a very similar ionic radius to magnesium. Consequently, $Fe^{2+}$ can replace the magnesium ion in the olivine structure.

One of the important factors that can increase the portion of forsterite in the olivine solid solution is the ratio of iron(II) ions to iron(III) ions in the magma. As the iron(II) ions oxidize and become iron(III) ions, iron(III) ions cannot form olivine because of their 3+ charge. The occurrence of forsterite due to the oxidation of iron was observed in the Stromboli volcano in Italy. As the volcano fractured, gases and volatiles escaped from the magma chamber. The crystallization temperature of the magma increased as the gases escaped. Because iron(II) ions were oxidized in the Stromboli magma, little iron(II) was available to form Fe-rich olivine (fayalite). Hence, the crystallizing olivine was Mg-rich, and igneous rocks rich in forsterite were formed.

### *Discovery and Name*

Forsterite was first described in 1824 for an occurrence at Mte. Somma, Vesuvius, Italy. It was named by Armand Lévy in 1824 after the English naturalist and mineral collector Jacob Forster.

### *Applications*

Forsterite is being currently studied as a potential biomaterial for implants owing to its superior mechanical properties.

## Fluorellestadite

Fluorellestadite is a rare nesosilicate of calcium, with sulphate and fluorine, with the chemical formula $Ca_{10}(SiO_4)_3(SO_4)_3F_2$. It is a member of the apatite group, and forms a series with hydroxylellestadite.

### *Etymology*

The mineral was originally named wilkeite by Eakle and Rogers in 1914, in honour of R. M. Wilke, a mineral collector and dealer. In 1922, a sample of "wilkeite" was analysed and found to be sufficiently different from the material reported by Eakle and Rogers to consider it a new species. The name "ellestadite" was proposed, in honour of Reuben B Ellestad (1900–1993), an American analytic chemist from the Laboratory for Rock Analysis, University of Minnesota, USA.

In 1982 Rouse and Dunn showed that the Si:S ratio was close to 1:1, giving the formula $Ca_{10}(SiO_4)_3(SO_4)_3X_2$, where X represents fluorine (F), hydroxyl (OH) or chlorine (Cl), and they named minerals in this group the ellestadite group. The end members of the group were named hydroxylellestadite (X = OH), fluorellestadite (X = F) and chlorellestadite (X = Cl); ideal end-member chlorellestadite is assumed not to exist in nature, although it has been synthesized. Wilkeite was discredited as a unique species, as it is not an end member of any solid solution series, but an intermediate member.

The name fluorellestadite was changed to ellestadite-(F) in 2008 and changed back to fluorellestadite in 2010.

### Structure

The ellestadites are nesosilicates, which are minerals with isolated $SiO_4$ tetrahedra. They are members of the apatite group, but whereas phosphorus is one of the chief constituents of apatite, in ellestadite it is almost completely replaced by sulfur and silicon, without appreciably altering the structure. The crystal class is hexagonal 6/m, space group $P6_3/m$. The tetrahedral groups are arranged to create the $6_3$ screw axis, and the fluorine atoms are located in channels parallel to this direction. Some sources give unit cell parametres for one formula unit per unit cell (Z = 1), but some scientists consider the formula to be half the value accepted by the International Mineralogical Association (IMA), i.e. $Ca_5((Si,S)O_4))_3F$, with two formula units per unit cell (Z = 2). Cell parametres for natural, as opposed to synthetic, material are a = 9.41 to 9.53 Å, and c = 6.90 to 6.94 Å. Rouse and Dunn postulated a hypothetical pure end-member with a = 9.543 Å and c = 6.917 Å. Synthetic material has a = 9.53 to 9.561 Å, and c = 6.91 to 6.920 Å.

### Appearance

Fluorellestadite occurs as acicular or hexagonal prismatic, poorly terminated crystals, and as fine-grained aggregates. Crystals are transparent and aggregates are translucent. Material from Crestmore, California, is light rose-red or yellow in colour, and typically occurs in a matrix of blue calcite. Material from Russia is pale bluish-green or colourless. The streak is white with a weak bluish tint, and the luster is sub-resinous on broken surfaces, but very brilliant on prism faces.

### Physical Properties

Fluorellestadite shows imperfect cleavage perpendicular to the long crystal axis. The mineral is very brittle, and breaks with a conchoidal fracture. Its hardness is 4½, between that of fluorite and apatite, and its specific gravity is 3.03 to 3.07, similar to that of fluorite. It is easily soluble in dilute hydrochloric and nitric acids and is not radioactive. When intensely heated, ellestadite (wilkeite) becomes colourless and then assumes a pale bluish green colour on cooling.

The mineral is uniaxial (-), with refractive indices $n_{\omega}$ = 1.638 to 1.655 and $n_{å}$ = 1.632 to 1.650. It is sometimes fluorescent, white to blue-white or yellow-white in short-wave ultraviolet light, and medium white-yellow-brown or weak white in long-wave light.

### *Occurrence and Associations*

The type locality is Coal Mine No. 44, Kopeisk, Chelyabinsk coal basin, Chelyabinsk Oblast, Southern Urals, Russia, and type material is held at the Fersman Mineralogical Museum, Academy of Sciences, Moscow, Russia. Ellestadite is a skarn mineral. It occurs associated with diopside, wollastonite, idocrase, monticellite, okenite, vesuvianite, calcite and others at Crestmore, Riverside County, California, USA. At Crestmore a contact zone exists between crystalline limestone and granodiorite. The area was quarried for limestone in the early 1900s, revealing varied associations of metamorphic minerals, including ellestadite (named as wilkeite) with garnet, vesuvianite and diopside, in blue calcite. At the type locality it was formed in burned fragments of petrified wood in coal dumps, associated with lime, periclase, magnesioferrite, hematite, srebrodolskite and anhydrite. Ellestadite (wilkeite) is often altered to okenite.

## Huttonite

Huttonite is a thorium nesosilicate mineral with the chemical formula $ThSiO_4$ and which crystallizes in the monoclinic system. It is dimorphous with tetragonal thorite, and isostructual with monazite. An uncommon mineral, huttonite forms transparent or translucent cream–coloured crystals. It was first identified in samples of beach sands from the West Coast region of New Zealand by the mineralogist Colin Osborne Hutton (1910–1971). Owing to its rarity, huttonite is not an industrially useful mineral.

### *Occurrence*

Huttonite was first described in 1950 from beach sand and fluvio-glacial deposits in South Westland, New Zealand, where it was found as anhedral grains of no more than 0.2 mm maximum dimension. It is most prevalent in the sand at Gillespie's Beach, near Fox Glacier, which is the type location, where it is accompanied by scheelite, cassiterite, zircon, uranothorite, ilmenite and gold. It was found at a further six nearby locations in less plentiful amounts. Huttonite was extracted from the sands by first fractionating in iodomethane and then electromagnetically. Pure samples were subsequently obtained by handpicking huttonite grains under a microscope. This was accomplished either in the presence of short wave (2540 Å) fluorescent light, where the dull white fluorescence distinguishes it from scheelite (fluoresces blue) and zircon (fluoresces yellow), or by first boiling the impure sample in hydrochloric acid to induce an oxide surface on scheelite and permitting handpicking under visible light.

Hutton suggested the huttonite contained in the beach sand and fluvio-glacial deposits originated from Otago schists or pegmatitic veins in the Southern Alps.

In addition to New Zealand, huttonite has been found in granitic pegmatites of Bogatynia, Poland, where it associated with cheralite, thorogummite, and ningyoite; and in nepheline syenites of Brevik, Norway.

### Physical Properties

Huttonite typically occurs as anhedral grains with no external crystal faces. It is usually colourless but also appears in colours; such as cream and pale yellow. It has a white streak. It has a hardness of 4.5 and exhibits distinct cleavage parallel to the c-axis [001] and an indistinct cleavage along the a-axis [100].

### Structure

Huttonite is a thorium nesosilicate with the chemical formula $ThSiO_4$. It is composed (by weight) of 71.59% thorium, 19.74% oxygen, and 8.67% silicon. Huttonite is found very close to its ideal stoichiometric composition, with impurities contributing less than 7% mole fraction. The most significant impurities to be observed are $UO_2$ and $P_2O_5$.

Huttonite crystallizes in the monoclinic system with space group $P2_1/n$. The unit cell contains four $ThSiO_4$ units, and has dimensions $a = 6.784 \pm 0.002$Å, $b = 6.974 \pm 0.003$Å, $c = 6.500 \pm 0.003$Å, and inter–axis angle $\beta = 104.92 \pm 0.03°$. The structure is that of a nesosilicate — discrete $SiO_4^{2-}$ tetrahedra coordinating thorium ions. Each thorium has coordination number nine. Axially, four oxygen atoms, representing the edges of two $SiO_4$ monomers on opposite sides of the thorium atom, form a $(–SiO_4–Th–)$ chain parallel to the *c* axis. Equatorially, five nearly planar oxygen atoms representing vertices of distinct silicate tetrahedra coordinate each thorium. The lengths of the axial Th–O bonds are 2.43 Å, 2.51 Å, 2.52 Å, 2.81 Å, and of the equatorial bonds, 2.40 Å, 2.41 Å, 2.41 Å, 2.50 Å, and 2.58 Å. The Si–O bonds are nearly equal, with lengths 1.58 Å, 1.62 Å, 1.63 Å, and 1.64 Å.

Huttonite is isostructural with monazite. Substitution of the rare earth elements and phosphorus of monazite with thorium and silicon of huttonite can occur to generates a solid solution. At the huttonite end-member, continuous rare earth substitution of thorium of up to 20% by weight has been observed. Thorium substitution in monazite has been observed up to 27% by weight. Substitution of $PO_4$ for $SiO_4$ also occurs associated with the introduction of fluoride, hydroxide, and metal ions.

Huttonite is dimorphic with thorite. Thorite crystallizes in a higher symmetry and lower density tetragonal form in which the thorium atoms coordinate to one less oxygen atom in an octahedral arrangement. Thorite is stable at lower temperatures than huttonite; at 1 atmosphere, the thorite–huttonite phase transition occurs between 1210 and 1225 °C. With increasing pressure the transition temperature increases. This relatively high transition temperature is thought to explain the relative rarity of huttonite on the Earth's crust. Unlike thorite, huttonite is not affected by metamictization.

## Jerrygibbsite

Jerrygibbsite is a rare silicate mineral with chemical formula: $(Mn,Zn)_9(SiO_4)_4(OH)_2$. Jerrygibbsite was originally discovered by Pete J. Dunn in 1984, who named it after mineralogist Gerald V. Gibbs (born 1929). It has only been reported from the type locality of Franklin Furnace, New Jersey, United States, and in Namibia's Otjozondjupa region. Jerrygibbsite is member of the leucophoenite family of the humite group. It is always found with these two minerals. It is a dimorph of sonolite.

### *Discovery*

The mineral jerrygibbsite was discovered in 1984 by Pete Dunn while conducting an X-ray spectrographic analysis of a sample previously assumed to be leucophoenicite. All samples found of jerrygibbsite are impure. All are incorporated within leucophoenicite, many by mixed layering, and tend to be found with many manganese humites such as sonolite. Physical properties are similar to those of leucophoenicite and sonolite, including hardness, colouring, and density.

### *Composition*

The formula for jerrygibbsite is $(Mn,Zn)_9(SiO_4)_4(OH)_2$, although it often contains impurities of iron, magnesium, calcium or water. The idealized formula is $Mn_9(SiO_4)_4(OH)_2$ which is the same ideal formula as sonolite, a member of the humite group. Jerrygibbsite has been found to be dimorphous with sonolite.

## Geologic Occurrence

Jerrygibbsite has been found only in the Franklin Furnace mine in Franklin, New Jersey, and in the Kombat Mines in Namibia. Most of the minerals in the humite group have been found only here, as well as leucophoenicite. Jerrygibbsite has been found to occur in contact with willemite, zincite, and sonolite in an uncommon assemblage. Jerrygibbsite typically occurs as a massive mineral in interlocking anhedral crystals, up to 0.5 mm × 2.0 mm, which display a typical metamorphic texture. Subsequent finds from the Namibia mines were of two different textures.

### *Physical Properties*

Jerrygibbsite, in pure form, is a violet-pink mineral with a light pink streak. It has a calculated density of 4.045 g/cm$^3$, and a tested density of 4.00 g/cm$^3$, agreeing favourably, since measurements used to test density have few significant figures. It has a hardness of about 5.5, that of a knife blade. The general luster is vitreous, or shiny. Crystals are generally transparent to translucent. Crystals are not luminescent or fluorescent. Jerrygibbsite forms orthorhombic crystals with an imperfect cleavage along the {001} plane, which can be seen by opaque lamellae alternating with the transparent jerrygibbsite. Optically, jerrygibbsite is negative biaxial with $2V = 72°$ and a maximum birefringence of 0.017. In thin section, jerrygibbsite appears light pink. The crystal structure described by Kato is the equivalent of a unit-cell-twinned sonolite in which the cells are related by a $b/4$ glide plane.

## Kyanite

Kyanite, whose name derives from the Greek word *kuanos* sometimes referred to as "kyanos", meaning deep blue, is a typically blue silicate mineral, commonly found in aluminium-rich metamorphic pegmatites and/or sedimentary rock. Kyanite in metamorphic rocks generally indicates pressures higher than four kilobars. Although potentially stable at lower pressure and low temperature, the activity of water is usually high enough under such conditions that it is replaced by hydrous aluminosilicates such as muscovite, pyrophyllite, or kaolinite. Kyanite is also known as disthene, rhaeticite and cyanite.

Kyanite is a member of the aluminosilicate series, which also includes the polymorph andalusite and the polymorph sillimanite. Kyanite is strongly anisotropic, in that its hardness varies depending on its crystallographic direction. In kyanite, this anisotropism can be considered an identifying characteristic.

At temperatures above 1100 °C kyanite decomposes into mullite and vitreous silica via the following reaction: $3(Al_2O_3 \cdot SiO_2) \rightarrow 3Al_2O_3 \cdot 2SiO_2 + SiO_2$. This transformation results in an expansion.

### *Uses*

Kyanite is used primarily in refractory and ceramic products, including porcelain plumbing fixtures and dishware. It is also used in electronics, electrical insulators and abrasives.

Kyanite has been used as a semiprecious gemstone, which may display cat's eye chatoyancy, though this use is limited by its anisotropism and perfect cleavage. Colour varieties include recently discovered orange kyanite from Tanzania. The orange colour is due to inclusion of small amounts of manganese (Mn3+) in the structure.

Kyanite is one of the index minerals that are used to estimate the temperature, depth, and pressure at which a rock undergoes metamorphism.

### *Notes for Identification*

Kyanite's elongated, columnar crystals are usually a good first indication of the mineral, as well as its colour (when the specimen is blue). Associated minerals are useful as well, especially the presence of the polymorphs of staurolite, which occur frequently with kyanite. However, the most useful characteristic in identifying kyanite is its anisotropism. If one suspects a specimen to be kyanite, verifying that it has two distinctly different hardness on perpendicular axes is a key to identification; it has a hardness of 5.5 parallel to {001} and 7 parallel to {100}.

### *Occurrence*

Kyanite occurs in gneiss, schist, pegmatite, and quartz veins resulting from high pressure regional metamorphism of principally pelitic rocks. It occurs as detrital grains in sedimentary rocks. It occurs associated with staurolite, andalusite, sillimanite, talc, hornblende, gedrite, mullite and corundum.

Kyanite occurs in Manhattan schist, formed under extreme pressure as a result of the two landmasses that formed supercontinent Pangaea.

## Majorite

Majorite is a type of garnet mineral found in the upper mantle of the Earth. Its chemical formula is $Mg_3(Fe,Al,Si)_2(SiO_4)_3$. It is distinguished from other garnets in having Si in octahedral as well as tetrahedral coordination. Majorite was first described in 1970 from the Coorara Meteorite of Western Australia and has been reported from various other meteorites in which majorite is thought to result from an extraterrestrial high pressure shock event. Mantle derived xenoliths containing majorite have been reported from potassic ultramafic magmas on Malaita Island on the Ontong Java Plateau Southwest Pacific.

### *Synthetic Magnesium Endmember Majorite*

Pure synthetic magnesium majorite ($MgSiO_3$)is a polymorph of enstatite, and akimotoite. Majorite is a member of the garnet group. It has Mg in eight-coordination with oxygen; it also has both Mg and Si in octahedral (6) coordination; and Si in tetrahedral (4) coordination with oxygen. Unlike most garnets, which are cubic, pure $MgSiO_3$ majorite is tetragonal.

### *Majorite in the Mantle*

Majorite is believed to be an abundant mineral in the lower transition zone and uppermost lower mantle of the Earth at depths of 550–900 kilometres (340–560 mi). It forms complex solid solutions with other Al, Fe, and Ca-bearing garnets in this region.

All of the minerals of the Earth's mantle are made of oxygen as the principal anion. It has been reported that a significant property of majorite is that under conditions of high pressure and temperature as exist in the mantle the mineral tends to absorb and store oxygen. However, when the temperature and pressure decrease as would occur when the majorite is drawn up towards the surface of the Earth by convection currents the mineral breaks down and releases the oxygen. Recent research has suggested that the total amount of oxygen stored in majorite in the mantle is likely quite large and may in fact contribute to keeping the Earth's surface moist and habitable.

## Malayaite

Malayaite is a calcium tin silicate mineral with formula $CaSnO[SiO_4]$. It is a member of the titanite group.

### *Discovery*

Malayaite was originally found in Perak (a state in Malaysia) and was first described in literature in 1961, though it was not yet given a name. In 1965, the mineral was named and recognized by the International Mineralogical Association. It was named for the locality in which it was discovered; which is the Malay Peninsula in Malaysia.

### Crystal Structure and Symmetry

This mineral is classified under the nesosilicate group for silicate minerals because the titanite group falls under this category as well. Minerals in the nesosilicate group have isolated $SiO_4$ tetrahedra connected to cations. In malayaite, the tetrahedra are connected to the chain of distorted $SnO_6$ octahedra, in which the octahedra are linked by vertex [*trans* corners] sharing and form chains parallel to the miller index of [100]. Within the $SnO_6$-$SiO_4$ framework, the $CaO_7$ polyhedra form chains parallel to [101].

Malayaite belongs to the monoclinic crystal system and has a 2/ m (prismatic) crystal class. According to the Hermann–Mauguin notation, the '2' refers to the two-fold axis while the 'm' refers to the presence of a single mirror plane. The '/' symbol indicates that the two fold axis is perpendicular to the mirror plane. The space group for malayaite is A2/a. According to the Bravais lattice symbol, 'A' refers to single face centering of the motif. This means that there is one more point on the centre of one face. '2' refers to the two-fold rotation axis while the '/' indicates that a mirror plane is perpendicular to the *a*-axis, which is the last symbol on that space group notation.

### Optical Properties

Malayaite exhibits anisotropy so it is doubly refractive and breaks up the light that travels through it into two distinct rays of different speeds. This mineral is known to have very high optical relief and has three indices of refraction. Since it is a monoclinic mineral, Malayaite exhibits two different colours under plane polarized light making it a pleochroic mineral. Malayaite is biaxial birefringent (trirefringent).

### Functions and Purposes

Malayaite has excellent thermal properties and stability making its host lattice ideal for ceramic pigmentation unlike its analogous mineral, titanite that undergoes phase transition at a higher temperature. This makes chromium-doped malayaite very desirable as it provides a consistent pigment for ceramics. Malayaite contains the element tin which is sometimes doped with chromium or nickel. These chromophore elements allows the mineral to produce different pigmentation. If malayaite is chromium-doped, it produces a pink-red colour while with nickel, it produces a purple colour. Pink chromium-doped malayaite $Ca(Sn,Cr)SiO_5$ is among a few important minerals used in producing pigments for the ceramic industry, specifically for colouring glazes as it produces a resilient colour.

### *Occurrence*

This mineral is generally found in tin-rich contact metamorphic skarn deposits. There is a possibility that malayaite is a hydrothermal altered form of cassiterite-quartz assemblage or some tin-bearing mineral. There have been specimens that were found with malayaite as a coating mineral on cassiterite.

## Mullite

Mullite or porcelainite is a rare silicate mineral of post-clay genesis. It can form two stoichiometric forms $3Al_2O_32SiO_2$ or $2Al_2O_3$ $SiO_2$. Unusually, mullite has no charge balancing cations present. As a result, there are three different Al sites: two distorted tetrahedral Al sites and one Al other site which adopts a higher co-ordinate octahedral state.

Mullite was first described in 1924 for an occurrence on the Isle of Mull, Scotland. It occurs as argillaceous inclusions in volcanic rocks in the Isle of Mull, inclusions in sillimanite within a tonalite at Val Sissone, Italy and with emerylike rocks in Sithean Sluaigh, Scotland.

### *Use in Porcelain*

Mullite is present in the form of needles in porcelain.

It is produced during various melting and firing processes, and is used as a refractory material, due to its high melting point of 1840°C.

In 2006 researchers at University College London and Cardiff University discovered that potters in the Hesse region of Germany since the late Middle Ages had used mullite in the manufacture of a type of crucible (known as Hessian crucibles), that were renowned for enabling alchemists to heat their crucibles to very high temperatures.

The formula for making it (using kaolinitic clay and then firing it at temperatures above 1100 °C) was kept a closely guarded secret.

Mullite morphology also is important for its application. In this case, there are two common morphologies for mullite. One is a platelet shape with low aspect ratio and the second is a needle shape with high aspect ratio. If the needle shape mullite can form in a ceramic body during sintering, it has an effect on both the mechanical and physical properties by increasing the mechanical strength and thermal shock resistance. The most important condition relates to ceramic chemical composition. If the silica and alumina ratio with low basic materials such as sodium and calcium is adjusted, the needle shape mullite forms at about 1400 °C and the needles will interlock. This mechanical interlocking causes the porcelain to have high mechanical strength.

### Use as a Catalyst

Recent research indicates that a synthetic analogue of mullite can be an effective replacement for platinum in reducing the amount of pollution generated by diesel engines.

## Olivine

The mineral olivine (when of gem quality, it is also called peridot and chrysolite) is a magnesium iron silicate with the formula ($Mg^{+2}$, $Fe^{+2}$)$_2SiO_4$. It is a common mineral in the Earth's subsurface but weathers quickly on the surface.

The ratio of magnesium and iron varies between the two endmembers of the solid solution series: forsterite (Mg-endmember: $Mg_2SiO_4$) and fayalite (Fe-endmember: $Fe_2SiO_4$). Compositions of olivine are commonly expressed as molar percentages of forsterite (Fo) and fayalite (Fa) (e.g., $Fo_{70}Fa_{30}$). Forsterite has an unusually high melting temperature at atmospheric pressure, almost 1900 °C, but the melting temperature of fayalite is much lower (about 1200 °C). The melting temperature varies smoothly between the two endmembers, as do other properties. Olivine incorporates only minor amounts of elements other than oxygen, silicon, magnesium and iron. Manganese and nickel commonly are the additional elements present in highest concentrations.

Olivine gives its name to the group of minerals with a related structure (the olivine group) which includes tephroite ($Mn_2SiO_4$), monticellite ($CaMgSiO_4$) and kirschsteinite ($CaFeSiO_4$).

Olivine's crystal structure incorporates aspects of the orthorhombic P Bravais lattice, which arise from each silica ($SiO_4$) unit being joined by metal divalent cations with each oxygen in $SiO_4$ bound to 3 metal ions. It has a spinel-like structure similar to magnetite but uses one quadravalent and two divalent cations $M_2^{+2}$ $M^{+4}O_4$ instead of two trivalent and one divalent cations.

### Identification and Paragenesis

Olivine is named for its typically olive-green colour (thought to be a result of traces of nickel), though it may alter to a reddish colour from the oxidation of iron.

Translucent olivine is sometimes used as a gemstone called peridot (*péridot*, the French word for olivine). It is also called chrysolite (or *chrysolithe*, from the Greek words for gold and stone). Some of the finest gem-quality olivine has been obtained from a body of mantle rocks on Zabargad island in the Red Sea.

Olivine/peridot occurs in both mafic and ultramafic igneous rocks and as a primary mineral in certain metamorphic rocks. Mg-rich olivine crystallizes from magma that is rich in magnesium and low in silica. That magma crystallizes to mafic rocks such as gabbro and basalt. Ultramafic rocks such as peridotite and dunite can be residues left after extraction of magmas, and typically they are more enriched in olivine after extraction of partial melts. Olivine and high pressure structural variants constitute over 50% of the Earth's upper mantle, and olivine is one of the Earth's most common minerals by volume. The metamorphism of impure dolomite or other sedimentary rocks with high magnesium and low silica content also produces Mg-rich olivine, or forsterite.

Fe-rich olivine is relatively much less common, but it occurs in igneous rocks in small amounts in rare granites and rhyolites, and extremely Fe-rich olivine can exist stably with quartz and tridymite. In contrast, Mg-rich olivine does not occur stably with silica minerals, as it would react with them to form orthopyroxene ($(Mg,Fe)_2Si_2O_6$).

Mg-rich olivine is stable to pressures equivalent to a depth of about 410 km within Earth. Because it is thought to be the most abundant mineral in Earth's mantle at shallower depths, the properties of olivine have a dominant influence upon the rheology of that part of Earth and hence upon the solid flow that drives plate tectonics. Experiments have documented that olivine at high pressures (e.g. 12 GPa, the pressure at depths of about 360 kilometres) can contain at least as much as about 8900 parts per million (weight) of water, and that such water contents drastically reduce the resistance of olivine to solid flow; moreover, because olivine is so abundant, more water may be dissolved in olivine of the mantle than contained in Earth's oceans.

## Extraterrestrial Occurrences

Mg-rich olivine has also been discovered in meteorites, the Moon, Mars, in the dust of comet Wild 2, within the core of comet Tempel 1, falling into infant stars, as well as on asteroid 25143 Itokawa. Such meteorites include chondrites, collections of debris from the early solar system; and pallasites, mixes of iron-nickel and olivine.

The spectral signature of olivine has been seen in the dust disks around young stars. The tails of comets (which formed from the dust disk around the young Sun) often have the spectral signature of olivine, and the presence of olivine has recently been verified in samples of a comet from the Stardust spacecraft. Comet-like (magnesium-rich) olivine has also been detected in the planetesimal belt around the star Beta Pictoris.

### Crystal Structure

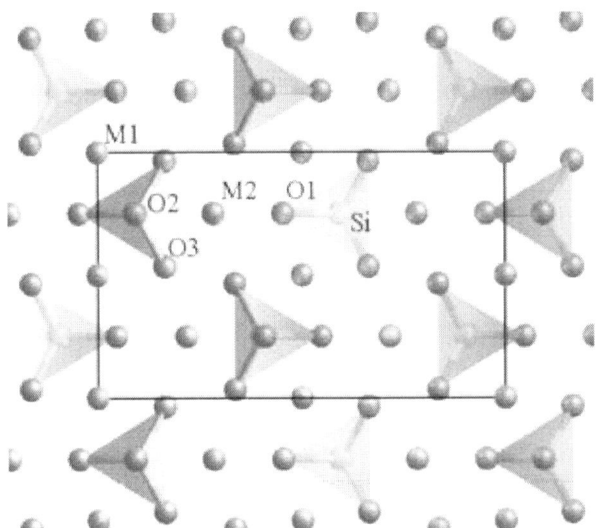


***Figure:*** *The atomic scale structure of olivine looking along the* a *axis. Oxygen is shown in red, silicon in pink, and magnesium/iron in blue. A projection of the unit cell is shown by the black rectangle*

Minerals in the olivine group crystallize in the orthorhombic system (space group P*bnm*) with isolated silicate tetrahedra, meaning that olivine is a nesosilicate. In an alternative view, the atomic structure can be described as a hexagonal, close-packed array of oxygen ions with half of the octahedral sites occupied with magnesium or iron ions and one-eighth of the tetrahedral sites occupied by silicon ions.

There are three distinct oxygen sites (marked O1, O2 and O3), two distinct metal sites (M1 and M2) and only one distinct silicon site. O1, O2, M2 and Si all lie on mirror planes, while M1 exists on an inversion centre. O3 lies in a general position.

### High Pressure Polymorphs

At the high temperatures and pressures found at depth within the Earth the olivine structure is no longer stable. Below depths of about 410 km (250 mi) olivine undergoes an exothermic phase transition to the sorosilicate, wadsleyite and, at about 520 km (320 mi) depth, wadsleyite transforms exothermically into ringwoodite, which has the spinel structure. At a depth of about 660 km (410 mi), ringwoodite decomposes into silicate perovskite ($(Mg,Fe)SiO_3$) and ferropericlase ($(Mg,Fe)O$) in an endothermic reaction. These phase transitions lead to a discontinuous increase in the density of the Earth's mantle that can be observed by seismic methods. They are also thought to influence the dynamics of mantle convection in that the exothermic transitions reinforce flow across the phase boundary, whereas the endothermic reaction hampers it.

The pressure at which these phase transitions occur depends on temperature and iron content. At 800 °C (1,070 K; 1,470 °F), the pure magnesium end member, forsterite, transforms to wadsleyite at 11.8 gigapascals (116,000 atm) and to ringwoodite at pressures above 14 GPa (138,000 atm). Increasing the iron content decreases the pressure of the phase transition and narrows the wadsleyite stability field. At about 0.8 mole fraction fayalite, olivine transforms directly to ringwoodite over the pressure range 10.0–11.5 GPa (99,000–113,000 atm). Fayalite transforms to $Fe_2SiO_4$ spinel at pressures below 5 GPa (49,000 atm). Increasing the temperature increases the pressure of these phase transitions.

### *Weathering*

Olivine is one of the weaker common minerals on the surface according to the Goldich dissolution series. It weathers to iddingsite (a combination of clay minerals, iron oxides and ferrihydrites) readily in the presence of water. The presence of iddingsite on Mars would suggest that liquid water once existed there, and might enable scientists to determine when there was last liquid water on the planet.

### *Uses*

A worldwide search is on for cheap processes to sequester $CO_2$ by mineral reactions, called enhanced weathering. Removal by reactions with olivine is an attractive option, because it is widely available and reacts easily with the (acid) $CO_2$ from the atmosphere. When olivine is crushed, it weathers completely within a few years, depending on the grain size. All the $CO_2$ that is produced by burning 1 litre of oil can be sequestered by less than 1 litre of olivine. The reaction is exothermic but slow. To recover the heat produced by the reaction to produce electricity, a large volume of olivine must be thermally well-isolated. The end-products of the reaction are silicon dioxide, magnesium carbonate and small amounts of iron oxide.

The aluminium foundry industry uses olivine sand to cast objects in aluminium. Olivine sand requires less water than silica sands while still holding the mold together during handling and pouring of the metal. Less water means less gas (steam) to vent from the mold as metal is poured into the mold.

In Finland, olivine is marketed as an ideal rock for sauna stoves because of its comparatively high density and resistance to erosion under repeated heating and cooling. Olivine is also used to tap blast furnaces in the steel industry, acting as a plug, removed in each steel run.

## Ringwoodite

Ringwoodite is the high-pressure polymorph of olivine that is stable at high temperatures and pressures of the Earth's mantle between 525 to 660 km depth. This mineral was first identified in the Tenham Meteorites in 1969, and it is inferred to be present in large quantity in the Earth's mantle. It was named after the Australian earth scientist Ted Ringwood (1930–1993) who studied polymorphic phase transitions in the common mantle minerals, olivine and pyroxene, at pressures equivalent to depths as great as about 600 km. Olivine, wadsleyite, and ringwoodite are polymorphs found in the upper mantle of the earth. At depths greater than about 660 km; other minerals, including some with the perovskite structure, are stable. The properties of these minerals determine many of the properties of the mantle.

Ringwoodite is the polymorph of olivine, $(Mg, Fe)_2SiO_4$, with the spinel structure. Spinel-group minerals crystallize in the isometric system with an octahedral habit. Olivine is most abundant in the upper mantle, above about 410 km; the olivine polymorphs, wadsleyite and ringwoodite, are thought to dominate the transition zone of the mantle, a zone present from about 410 to 660 km depth. Ringwoodite is thought to be the most abundant mineral phase in the lower part of Earth's transition zone. The physical and chemical property of this mineral partly determine properties of the mantle at those depths. The pressure range for stability of ringwoodite lies in the approximate range from 18 to 23 GPa.

Apart from the mantle, natural ringwoodite has been found in many shocked chondritic meteorites, in which the ringwoodite occurs as fine-grained polycrystalline aggregates.

### *Geological Occurrences*

In meteorites, ringwoodite occurs in the veinlets of quenched shock-melt cutting the matrix and replacing olivine probably produced during shock metamorphism. In Earth's interior, olivine occurs in the upper mantle at depths less than about 410 km, and ringwoodite is inferred to be present within the transition zone from about 520 to 660 km depth. Seismic discontinuities at about 410, 520, and 660 km depth have been attributed to phase changes involving olivine and its polymorphs. The 520-km discontinuity is generally believed to be caused by the transition of the olivine polymorph, wadsleyite (beta-phase) to ringwoodite (gamma-phase), while the 660-km discontinuity by the phase transformation of ringwoodite (gamma-phase) to a silicate perovskite plus magnesiowüstite.

Ringwoodite in the lower half of the transition zone is inferred to play a pivotal role in mantle dynamics, and the plastic properties of ringwoodite are thought to be critical in determining flow of material in this part of the mantle. The solubility of hydroxide in ringwoodite is important because of the effect of hydrogen upon rheology. Ringwoodite synthesized at conditions appropriate for the transition zone has been found to contain up to 2.6 weight percent water. Because the transition zone between the Earth's upper and lower mantle helps govern the scale of mass and heat transport throughout the Earth, the presence of water within this region, whether global or localized, may have a significant effect on mantle rheology and therefore mantle circulation. In regions of subduction zones, the ringwoodite stability field hosts high levels of seismicity.

### *Crystal Structure*

Ringwoodite crystallizes in the isometric crystal system with space group *Fd*3*m*. On the atomic scale, magnesium and silicon are in octahedral and tetrahedral coordination with oxygen, respectively. The Si-O and Mg-O bonds are both ionic and covalent. The cubic unit cell parametre is 8.063 Å for pure $Mg_2SiO_4$ and 8.234 Å for pure $Fe_2SiO_4$.

### *Chemical Composition*

Ringwoodite compositions range from pure $Mg_2SiO_4$ to $Fe_2SiO_4$ in synthesis experiments. Ringwoodite can incorporate up to 2.6 percent by weight $H_2O$ .

### *Physical Properties*

The physical properties of ringwoodite are affected by the pressure and temperature. The calculated density value of ringwoodite is 3.564

g/cm³ for pure $Mg_2SiO_4$; 3.691 for Fo90 composition of typical mantle; and 4.845 for $Fe_2SiO_4$. It is an isotropic mineral with an index of refraction n = 1.768.

The colour of ringwoodite varies between the meteorites, between different ringwoodite bearing aggregates, and even in one single aggregate. The ringwoodite aggregates can show every shade of blue, purple, grey and green, or they have no colour at all. A closer look at coloured aggregates shows that the colour is not homogeneous, but seems to originate from something with a size similar to the ringwoodite crystallites. In synthetic samples, pure Mg ringwoodite is colourless, whereas samples containing more that one mole percent $Fe_2SiO_4$ are deep blue in colour. The colour is thought to be due to $Fe^{2+}$-$Fe^{3+}$ charge transfer.

## Topaz

Topaz is a silicate mineral of aluminium and fluorine with the chemical formula $Al_2SiO_4(F,OH)_2$. Topaz crystallizes in the orthorhombic system, and its crystals are mostly prismatic terminated by pyramidal and other faces.

### *Colour and Varieties*

Pure topaz is colourless and transparent but is usually tinted by impurities; typical topaz is wine, yellow, pale gray, reddish-orange, or blue brown. It can also be made white, pale green, blue, gold, pink (rare), reddish-yellow or opaque to transparent/translucent.

*Orange topaz,* also known as precious topaz, is the traditional November birthstone, the symbol of friendship, and the state gemstone of the US state of Utah.

*Imperial topaz* is yellow, pink (rare, if natural) or pink-orange. Brazilian Imperial Topaz can often have a bright yellow to deep golden brown hue, sometimes even violet. Many brown or pale topazes are treated to make them bright yellow, gold, pink or violet coloured. Some imperial topaz stones can fade on exposure to sunlight for an extended period of time.

*Blue topaz* is the state gemstone of the US state of Texas. Naturally occurring blue topaz is quite rare. Typically, colourless, gray or pale yellow and blue material is heat treated and irradiated to produce a more desired darker blue.

*Mystic topaz* is colourless topaz which has been artificially coated giving it the desired rainbow effect.

### Localities and Occurrence

Topaz is commonly associated with silicic igneous rocks of the granite and rhyolite type. It typically crystallizes in granitic pegmatites or in vapour cavities in rhyolite lava flows like those at Topaz Mountain in western Utah. It can be found with fluorite and cassiterite in various areas including the Ural and Ilmen mountains of Russia, in Afghanistan, Sri Lanka, Czech Republic, Germany, Norway, Pakistan, Italy, Sweden, Japan, Brazil, Mexico; Flinders Island, Australia; Nigeria and the United States.

Some clear topaz crystals from Brazilian pegmatites can reach boulder size and weigh hundreds of pounds. Crystals of this size may be seen in museum collections. The Topaz of Aurangzeb, observed by Jean Baptiste Tavernier measured 157.75 carats. The American Golden Topaz, a more recent gem, measured a massive 22,892.5 carats.

Colourless and light-blue varieties of topaz are found in Precambrian granite in Mason County, Texas within the Llano Uplift. There is no commercial mining of topaz in that area.

### Etymology

The name "topaz" is derived (via Old French: Topace and Latin: Topazus) from the Greek Ôïpáziïs or Ôïpáziïn, the ancient name of St. John's Island in the Red Sea which was difficult to find and from which a yellow stone (now believed to be chrysolite: yellowish olivine) was mined in ancient times; topaz itself (rather than *topazios*) was not really known about before the classical era. Pliny said that Topazos is a *legendary* island in the Red Sea and the mineral "topaz" was first mined there.

The word *topaz* is related to the Sanskrit word "tapas" meaning "heat" or "fire", and also to the Hebrew word for "orange" (the fruit): *tapooz* (úôåæ), both of which predate the Greek word.

### Historical Usage

Nicols, the author of one of the first systematic treatises on minerals and gemstones, dedicated two chapters to the topic in 1652. In the Middle Ages, the name topaz was used to refer to any yellow gemstone, but in modern times it denotes only the silicate described above.

Many modern English translations of the Bible, including the King James Version mention *topaz* in Exodus 28:17 in reference to a stone in the Hoshen: "And thou shalt set in it settings of stones, even four rows of stones: the first row shall be a sardius, a topaz, and a carbuncle (garnet): this shall be the first row."

However, because these translations as *topaz* all derive from the Septuagint translation *topazi[os]*, which as mentioned above referred to a yellow stone that was not topaz, but probably *chrysolite* (chrysoberyl or peridot), it should be borne in mind that topaz is likely not meant here. The masoretic text (the Hebrew on which most modern Protestant Bible translations of the Old Testament are based) has *pitdah* as the gem the stone is made from; some scholars think it is related to an Assyrian word meaning "flashed".

More likely, pitdah is derived from Sanskrit words (pit = yellow, dah = burn), meaning "yellow burn" or, metaphorically, "fiery".

## Tranquillityite

Tranquillityite is silicate mineral with formula $(Fe^{2+})_8Ti_3Zr_2Si_3O_{24}$. It is mostly composed of iron, oxygen, silicon, zirconium and titanium with smaller fractions of yttrium and calcium. It is named after the Mare Tranquillitatis (Sea of Tranquility), the place on the Moon from which the rock samples in which it was found were brought during the Apollo 11 mission in 1969. Until its discovery in Australia in 2011, it was the last mineral brought from the Moon which was thought to be unique, with no terrestrial counterpart.

### *Discovery*

In 1970, material scientists found a new unnamed Fe, Ti, Zr-silicate mineral containing rare-earths and Y in lunar rock sample 10047. The first detailed analysis of the mineral was published in 1971 and the name "tranquillityite" was proposed and later accepted by the International Mineralogical Association. It was later found in lunar rock samples from all Apollo missions. Samples were dated by Pb/Pb ion probe techniques. Together with armalcolite and pyroxferroite, it is one of the three minerals which were first discovered on the Moon, before terrestrial occurrences were found. Fragments of tranquillityite were later found in Northwest Africa, in the NWA 856 Martian meteorite.

Terrestrial occurrences of tranquillityite have been found in six localities in the Pilbara region of Western Australia, Western Australia in 2011. The Australian occurrences include a number of Proterozoic to Cambrian age diabase and gabbro dikes and sills. It occurs as interstitial grains with zirconolite, baddeleyite, and apatite associated with late stage intergrowths of quartz and feldspar.

### *Properties*

Tranquillityite forms thin stripes up to 15 by 65 micrometres in size in basaltic rocks, where it was produced at a late crystallization stage. It is associated with troilite, pyroxferroite, cristobalite and alkali feldspar. The mineral is nearly opaque and appears dark red-brown in thin crystals. The analyzed samples contains less than 10% impurities (Y, Al, Mn, Cr, Nb and other rare-earth element) and up to 0.01% (100 ppm) of uranium. Presence of significant amount of uranium allowed to estimate the age of tranquillityite and some associated minerals in Apollo 11 samples as 3710 million years using the uranium-lead dating technique.

Irradiation by alpha particles generated by the uranium decay is believed to be the origin of the predominantly amorphous metamict structure of tranquillityite. Its crystals were obtained by annealing the samples at 800 °C (1,470 °F) for 30 minutes. Longer annealing did not improve the crystalline quality, and annealing at higher temperatures resulted in spontaneous fracture of samples.

The crystals were initially found to have a hexagonal crystal structure with the lattice parametres, $a = 1.169$ nm, $c = 2.225$ nm and three formula units per unit cell, but later reassigned a face-centred cubic structure (fluorite-like). A tranquillityite-like crystalline phase has been synthesized by mixing oxide powders in an appropriate ratio, determined from the chemical analysis of the lunar samples, and annealing the mixture at 1,500 °C (2,730 °F). This phase was not pure, but intergrown with various intermetallic compounds.

# 6

# Sorosilicates

## Åkermanite

Åkermanite ($Ca_2Mg[Si_2O_7]$) is a melilite mineral of the sorosilicate group, containing calcium, magnesium, silicon, and oxygen. It is a product of contact metamorphism of siliceous limestones and dolostones, and rocks of sanidinite facies. Sanidinite facies represent the highest conditions of temperature of contact metamorphism and are characterized by the absence of hydrous minerals. It has a density of 2.944 g/cm$^3$. Åkermanite ranks a 5 or 6 on the Mohs scale of mineral hardness, and can be found grey, green, brown, or colourless. It has a white streak and a vitreous or resinous luster. It has a tetragonal crystal system and a good, or distinct, cleavage. It is the end member in a solid solution series beginning with gehlenite ($Ca_2Al[AlSiO_2]$).

The mineral is named for Anders Richard Åkerman (1837–1922), a Swedish metallurgist. It has been found at Monte Somma and Vesuvius, and Monte Cavalluccio near Rome. It was "grandfathered" in as a species of mineral because it was described prior to 1959, before the founding of the International Mineralogical Association.

### *Axinite*

Axinite is a brown to violet-brown, or reddish-brown bladed group of minerals composed of calcium aluminium boro-silicate, $(Ca,Fe,Mn)_3Al_2BO_3Si_4O_{12}OH$. Axinite is pyroelectric and piezoelectric.

The axinite group includes:

- Axinite-(Fe) or ferroaxinite, $Ca_2Fe^{2+}Al_2BOSi_4O_{15}(OH)$ iron rich, clove-brown, brown, plum-blue, pearl-gray

- Axinite-(Mg) or magnesioaxinite, $Ca_2MgAl_2BOSi_4O_{15}(OH)$ magnesium rich, pale blue to pale violet; light brown to light pink
- Axinite-(Mn) or manganaxinite, $Ca_2Mn^{2+}Al_2BOSi_4O_{15}(OH)$ manganese rich, honey-yellow, clove-brown, brown to blue
- Tinzenite, $(CaFe^{2+}Mn^{2+})_3Al_2BOSi_4O_{15}(OH)$ iron – manganese intermediate, yellow, brownish yellow-green

### *Cervandonite*

Cervandonite is a rare arsenosilicate mineral. It has a chemical formula $(Ce,Nd,La)(Fe^{3+},Fe^{2+},Ti^{4+},Al)_3SiAs(Si,As)O_{13}$ or (Ce,Nd,La) $(Fe^{3+},Fe^{2+},Ti,Al)_3O_2\,(Si_2O_7)\,(As^{3+}O_3)$ (OH). It has a monoclinic crustal structure with supercell (Z=6), the crystal structure was established as a trigonal subcell, with space group R3m and a = 6.508(1)Å, c = 18.520(3) Å, V 679.4(2) $Å^3$, and Z=3. It was first described by Buhler Armbruster in 1988, but it has proven to be problem due to the extreme scarcity of single crystals and its unusual replacement of silicon and arsenic. Cervandonite is named after the location where it was first described, Pizzo Cervandone (Scherbadung), Italy in the Central Alps.

### *Structure*

The current work is based on a single- crystal fragment of cervandonite measuring 0.08 X 0.03 X 0.02 mm. The exact fragment was re-examined with MoKá radiation using BRUKER Apex II diffractometre equipped with a 2K CCD detector. Through the use of X-ray diffraction, the unusual nature of the As to Si substitution found arsenic to be present as $As^{3+}$ instead of $As^{5+}$, with the presence of sorosilicate $Si_20_7^{6-}$ anions were also established. The original description of the cervandonite mineral, (Ce,Nd,La) $(Fe^{3+},Fe^{2+},Ti^{4+},Al)_3$ SiAs $(Si,As)O_{13}$, was rewritten to, (Ce,Nd,La) $(Fe^{3+},Fe^{2+},Ti^{4+},Al)_3\,O_2(Si_2As_7)_{1-x+y}(As0_3)_{1+x-y}(OH)_{3x-3y}$. The x and y values are 0.47 and 0.31. Even though the values were derived from a refined and disordered crystal, the new formula matches the charge balance [13 negative versus 12.75 positive charges] for the average chemical composition for the M(1)+M(2) sites. The As-containing silicates, none of which contained the $Si_2As_7$ composition, or arsenic in the 3+ state, where considered to be examples of As to Si diadochy. However, after accurate determination of the crystal structure the As and Si atoms where shown to occupy distinct cell sites. So far, the As-Si disorder is unique for the cervandonite mineral, and accounts for the As:Si ratio. Due to the unusual diadochy, the variable composition of cervandonite might consist of an assembly of more or less twinned microdomains. Since cervandonite contains different values of As/Si and unit- cell parametres, has observed one type of superstructure.

## Physical Properties

Cervandonite- (Ce) has a monoclinic cell with Z=6. The monoclinic structure was refined as a trigonal subcell using 411 reflections with $I > 2\sigma (I)$, R1=0.320, wR2=0.0887. The R- centred cell can be transformed with, a 6.508 (1), c 18.520 (3) Å, V 679.4(2) ú3, and Z=3. The structure has a space group of R3m, the solution of the cervandonite structure instantly revealed the presence of one eight-coordinated structure, REE-containing M(1) lying on the threefold axis and M(2) which includes Fe, Ti, or Al. The colour of the mineral is a black, with transmitted light it will reflect a yellowish, reddish brown to black colour. The mineral is brittle, porous, rosettelike aggregate, with adamantine luster, poorest at {001} cleavage, conchoidal fracture with a brownish black streak, and a hardness= 5.0. Although cervandonite was found in Pizzo Cervandone, it is not the only rare mineral discovered there, much like the crystal fetiasite, which shares common morphology, fetiasite has a thin brown-red alteration layer with the perfect cleavage is on {100}. Cervandonite was discovered in the east region of Pizzo Cervandone, Alpe Devero, on the border of Italy and Switzerland, and on the west region of Cherbadung, Switzerland. This mountain is well known for Alpine excursionists and mineral collectors. It was known the 1960s as a site for rare and new minerals.

### *Ericssonite*

Ericssonite has a general formula of $BaMn_2FeO[Si_2O_7](OH)$. It was discovered in 1967 and named after John Ericsson (July 31, 1803 – March 8, 1889), a well known Swedish American inventor, engineer and designer of the iron-clad ship USS *Monitor*. Ericssonite was discovered in the Jakobsberg Mine in Värmland, Sweden. Ericcsonite is monoclinic; this means it contains three unequal vectors, two of these vectors angles are perpendicular while the other is at an angle greater than 90°. When talking about its optical properties, ericssonite is anisotropic which means that the mineral has more than 1 index of refraction, causing light to vary in speed depending on which axis it is travelling through. The value of relief, the way the mineral appears to stand out when viewing it in plane polarized(PP) light under a microscope, in this mineral ranges anywhere from 1.802-1.891. This means that light travels anywhere from 1.802-1.891 times as fast in this mineral as it does in air, assuming the index of refraction of light in a vacuum is one. Since ericssonite is monoclinic, containing three unequal vectors, it is no surprise that it has three index of refraction, which is just the measurement of the speed of light in the mineral compared to the speed of light in a vacuum. Lastly when looking at ericssonite in PP light it is usually seen as a deep reddish-black. Ericssonite is only found in the Langban mine in Sweden, and usually in a metamorphic manganese ore body. Also it is always intergrown with orthoericssonite, which is almost identical to ericssonite except you will find an extra silicon and oxygen in its chemical formula. Besides the fact that it is magnetic, the most unique aspect of ericssonite is that it is just a very rare mineral, and it is only found in one place in the world, making it most useful for a collector of rare minerals.

## Gehlenite

Gehlenite, ($Ca_2Al[AlSiO_7]$), is a sorosilicate, Al-rich endmember of the melilite complete solid solution series with akermanite. The type locality is in the Monzoni Mountains, Fassa Valley in Trentino in Italy, and is named after Adolf Ferdinand Gehlen (1775–1815) by A.J. Fuchs in 1815.

### *Geological Occurrence*

Gehlenite is found in carbonaceous chondrites from which it condensed as a refractory mineral in the hotter stages (FU Ori) of the presolar nebula, and was subsequently consumed in processes which created enstatite and other more abundant minerals making it a remanent (badword) mineral from the early solar nebula (along with corrundum and spinel). Its occurrence (badword) in the early condensation phase of the solar nebula was predicted by Harry Lord in the 1950s, but studies of carbonaceous chondrites did not support this claim until the Allende meteorite was discovered in 1969. It is also found in diorite intruded carbonate rocks, and to a far lesser extent in uncompahgrites, melilitites, alnoites, lamprophyres and possibly kimberlite pipes.

Gehlenite has also been found on the comet 81P/Wild.

### *Crystallography, Composition and Physical Properties*

Gehlenite is one of five, isostructural tetragonal crystal system minerals in the melilite group. The tetrahedral linkage within the structure is similar to that of an aluminosilicate framework structure and was once considered a feldspathoid-like mineral due to silica undersaturation.

Gehlenite has a Mohs hardness of 5-6, a vitreous to greasy lustre, distinct to good cleavage and is yellow brown, greenish grey or colourless. Its streak is white or grey-white. It is uniaxial (-), has an anomalous nonzero 2V angle and has a characteristic 'ultrablue' birefringence.

## Hardystonite

Hardystonite is a rare calcium zinc silicate mineral first described from the Franklin, New Jersey, USA zinc deposits. It often contains lead, which was detrimental to the zinc smelting process, so it was not a useful ore mineral. Like many of the famous Franklin minerals, hardystonite responds to short wave ultraviolet (254 nm wavelength) light, emitting a fluorescence from dark purple to bright violet blue. In daylight, it is white to gray to light pink in colour, sometimes with a vitreous or greasy luster. It is very rarely found as well formed crystals, and these are usually rectangular in appearance and rock-locked.

Hardystonite has a chemical composition of $Ca_2ZnSi_2O_7$. It is frequently found with willemite (fluoresces green), calcite (fluoresces red), and clinohedrite (fluoresces orange). Hardystonite can be found altered to clinohedrite $CaZn(SiO_4)\cdot H_2O$ through direct hydrothermal alteration. Other minerals often associated with hardystonite are franklinite, diopside, andradite garnet, and esperite (fluoresces yellow).

It was first described in 1899 by J.E. Wolff, when the New Jersey Zinc Company mines were located in what was called Franklin Furnace, in Hardyston Township, New Jersey.

### *Hemimorphite*

Hemimorphite, is a sorosilicate mineral which has been mined from days of old from the upper parts of zinc and lead ores, chiefly associated with smithsonite. It was often assumed to be the same mineral and both were classed under the same name of calamine. In the second half of the 18th century it was discovered that there were two different minerals under the heading of calamine - a zinc carbonate and a zinc silicate, which often closely resembled each other.

The silicate was the more rare of the two, and was named hemimorphite because of the hemimorph development of its crystals. This unusual form, which is typical of only a few minerals, means that the crystals are terminated by dissimilar faces. Hemimorphite most commonly forms crystalline crusts and layers, also massive, granular, rounded and reniform aggregates, concentrically striated, or finely needle-shaped, fibrous or stalactitic, and rarely fan-shaped clusters of crystals.

Some specimens show strong green fluorescence in shortwave ultraviolet light (253.7 nm) and weak light pink fluorescence in longwave UV.

### *Occurrence*

Hemimorphite most frequently occurs as the product of the oxidation of the upper parts of sphalerite bearing ore bodies, accompanied by other secondary minerals which form the so-called *iron cap* or *gossan*. Hemimorphite is an important ore of zinc and contains up to 54.2% of the metal.

The regions on the Belgian-German border are well known for their deposits of hemimorphite of metasomatic origin, especially Vieille Montagne in Belgium and Aachen in Germany. Other deposits are near Tarnovice in upper Silesia, Poland; near Phoenixville, Pennsylvania; the Missouri lead-zinc district; Elkhorn, Montana; Leadville, Colorado; and Organ Mountains, New Mexico in the United States; and in several localities in North Africa. Further hemimorphite occurrences are the Padaeng deposit near Mae Sod in western Thailand; Sardinia; Nerchinsk, Siberia; Cave del Predil, Italy; Bleiberg, Carinthia, Austria; Matlock, Derbyshire, England.

## Julgoldite

Julgoldite is a member of the pumpellyite mineral series, a series of minerals characterized by the chemical bonding of silica tetrahedra with alkali and transition metal cations. Julgoldites, along with more common minerals like epidote and vesuvianite, belong to the subclass of sorosilicates, the rock-forming minerals that contain $SiO_4$ tetrahedra that share a common oxygen to form $Si_2O_7$ ions with a charge of 6- (Deer et al., 1996). Julgoldite has been recognized for its importance in low grade metamorphism, forming under shear stress accompanied by relatively low temperatures (Coombs, 1953). Julgoldite was named in honour of Professor Julian Royce Goldsmith (1918–1999) of the University of Chicago.

### *Composition*

The chemical formula of julgoldite is $(Ca,Mn)_2$ $(Fe^{2+},Fe^{3+},Mg)$ $(Fe^{3+},Al)_2$ $(SiO_4)$ $(Si_2O_7)$ $(OH)_2$ $(H_2O)$ (Moore, 1971). Pumpellyites are classified according to the prevailing metals in the X and Y sites (Moore, 1971). When Mg in the X position and Al in the Y position and both occupy greater than 50% molarity of their positions, then the mineral is identified as a pumpellyite (Moore, 1971). Julgoldites are identified when $Fe^{2+}$ in the X and $Fe^{3+}$ in the Y each occupy greater than 50% molarity of their positions (Moore, 1971).

## Geologic Occurrence

Julgoldites were first collected as samples entrenched in large plates of apophyllite and barite, comprising a fissure inside granular hematite-magnetite ore in Långban, Sweden (Moore, 1971). Julgoldite has since been discovered in other parts of the world: Edinburgh, Scotland (in quartz dolerite) (Livingstone, 1976) and Norilsk, Taymyr Peninsula, Russia, one of the largest nickel deposits in the world, in metamorphosed basalts and diabases associated with prehnite and laumontite (Zolotukin et al., 1965). Julgoldite has also been found exposed in basalt cavities in the Khondivili Quarry near Bombay, India along with other silicates, including pumpellyite-$Fe^{2+}$, ilvaite, babingtonite, hydroandradite, prehnite, and chlorite (Wise and Moller, 1990). These minerals crystallized in the same basaltic cavities, which were primarily formed from gas bubbles in the compound lava flows; all of these Ca-Fe silicates formed in different phases of a low temperature environment (Wise and Moller, 1990).

### *Atomic Structure*

The atomic structures of pumpellyites and julgoldites consist of chains of edge-sharing octahedra linked by $SiO_4$, $Si_2O_7$, and $CaO_7$ polyhedra: in the julgoldite atomic structure, the Ca site, the W site, is a seven-coordinated site with oxygen; the X and Y are two crystallographically independent octahedral sites; and the $SiO_4$ site is tetrahedral (Passaglia and Gottardi, 1973). Like epidote, for which the chemical formula is $Ca_2(Fe^{3+},Al)Al_2(SiO_4)(Si_2O_7)(OH)$, julgoldite contains additional $SiO_4$ tetrahedra that are independent of the $Si_2O_7$ structural units (Deer et al., 1996). The octahedral sites form chains along the b axis by sharing opposite edges (Allman and Donnay, 1973). The octahedral chains are joined in the ac plane by $SiO_4$ and $Si_2O_6$ (OH) groups, forming five-member rings of two octahedra and three tetrahedral (Allman and Donnay, 1973). Half of the rings are open ended and have a $Ca^{2+}$ ion in their centre; the other rings are closed, and they surround a Ca2+ ion (Allman and Donnay, 1973). The X and Y chains are parallel to the crystallographic direction [010]; therefore, the two edge sharing polyhedra cause variations in the b cell parametre (Artioli et al., 2003). Two layers of X chains and one layer of Y chains occur along the [100] direction, whereas two layers of both X and Y chains occur along the [001] direction (Artioli et al., 2003).

### *Physical Properties*

Moore (1971) sampled flat prismatic or bladed crystals, with the greatest dimension of each mineral to be no more than 2mm. The julgoldites found in the basalt cavities in India were almost 10 mm in length (Wise and Moller, 1990). Julgoldite is elongated parallel b [010] and flattened parallel a {100} (Moore, 1971). Allman and Donnay (1973) calculated the cell dimensions to be a 8.922(4), b 6.081(3), c 19.432 (9) Å. The colour of julgoldite is usually a deep lustrous black, and it has a hardness of 4.5 and cleavage on the a-axis {100} (Moore, 1971). It has a greenish-olive powdery streak with a blue tinge (Moore, 1971). Under the petrographic microscope, a thin section of the mineral will display brilliant interference figures in greens or blues (Moore, 1971). The mineral is classified under the space group A2/m (Moore, 1971). A monoclinic mineral, julgoldite is isostructural to pumpellyite and epidote (Allman and Donnay, 1973).

## Junitoite

Junitoite is a mineral with formula $CaZn_2Si_2O_7 \cdot H_2O$. It was discovered at the Christmas mine in Christmas, Arizona, and described in 1976. The mineral is named for mineral chemist Jun Ito (1926–1978).

### *Description and Occurrence*

Junitoite is transparent to translucent and is colourless, milk-white, or coloured due to alteration. Crystals grow up to 5 millimetres (0.20 in) and have high quality faces.

Junitoite occurs in fractures through pods of sphalerite. It formed by retrograde metamorphism and oxidation of tactite, also resulting in kinoite. The mineral is known from New Jersey and the type locality in Arizona. Junitoite occurs in association with apophyllite, calcite, kinoite, smectite, and xonotlite.

### *Crystal Structure*

In 1968, Jun Ito published the results of synthesis of various lead calcium zinc silicates. The formula of one phase, designated $X_3$, was identified as probably $CaZnSi_2O_6 \cdot H_2O$. When he described junitoite, Sidney Williams identified the mineral's formula as $CaZn_2Si_2O_7 \cdot H_2O$, based on communications with Ito.

The mineral's crystal structure was first determined in 1985 and refined in 2012. The mineral crystallizes in the orthorhombic crystal system. The structure is formed by chains of corner-sharing $ZnO_4$ tetrahedra linked together by $Si_2O_7$ tetrahedral pairs. Calcium ions occupy vacancies and coordinate to five oxygen atoms and one water molecule.

### History

The first known specimen of junitoite was collected from the Christmas mine at Christmas, Arizona, and entered the collection of Joe Ana Ruiz. Geologist Robert A. Jenkins noticed the mineral in kinoite specimens, submitting Ruiz's sample to Sidney A. Williams for study. Further samples came from the collections of Ruiz and Raymond Diaz.

Williams identified the specimens as a new mineral and described it in the journal *American Mineralogist* in 1976. He named it *junitoite* in honour of Jun Ito, the mineral chemist who noted the compound of which the mineral is composed. The International Mineralogical Association approved the mineral as IMA 1975-042. The type material is housed in the University of Arizona, Harvard University, the National Museum of Natural History, the University of Paris, the National School of Mines, and The Natural History Museum.

## Pumpellyite

Pumpellyite is a group of closely related sorosilicate minerals:

- pumpellyite-(Mg): $Ca_2MgAl_2[(OH)_2 | SiO_4 | Si_2O_7] \cdot (H_2O)$
- pumpellyite-(Fe2+): $Ca_2Fe^{2+}Al_2[(OH)_2 | SiO_4 | Si_2O_7] \cdot (H_2O)$
- pumpellyite-(Fe3+): $Ca_2(Fe^{3+},Mg,Fe^{2+}) (Al,Fe^{3+})_2 [(OH,O)_2 | SiO_4 | Si_2O_7] \cdot H_2O$
- pumpellyite-(Mn2+): $Ca_2(Mn^{2+},Mg)(Al,Mn^{3+},Fe^{3+})_2[(OH)_2 | SiO_4 | Si_2O_7] \cdot (H_2O)$
- pumpellyite-(Al): $Ca_2(Al,Fe^{2+},Mg)Al_2[(OH,O)_2 | SiO_4 | Si_2O_7] \cdot H_2O$

Pumpellyite crystallizes in the monoclinic-prismatic crystal system. It typically occurs as blue-green to olive green fibrous to lamellar masses. It is translucent and glassy with a Mohs hardness of 5.5 and a specific gravity of 3.2. It has refractive indices of $n_\alpha$=1.674–1.748, $n_\beta$=1.675–1.754 and $n_\gamma$=1.688–1.764.

Pumpellyite occurs as amygdaloidal and fracture fillings in basaltic and gabbroic rocks in metamorphic terranes. It is an indicator mineral of the prehnite-pumpellyite metamorphic facies. It is associated with chlorite, epidote, quartz, calcite and prehnite. It was first described in 1925 for occurrences in the Calumet mine, Houghton Co., Keweenaw Peninisula, Michigan and named for United States geologist, Raphael Pumpelly (1837–1923).

### *Lawsonite*

Lawsonite is a hydrous calcium aluminium sorosilicate mineral with formula $CaAl_2Si_2O_7(OH)_2 \cdot H_2O$. Lawsonite crystallizes in the orthorhombic system in prismatic, often tabular crystals. Crystal twinning is common. It forms transparent to translucent colourless, white, and bluish to pinkish grey glassy to greasy crystals. Refractive indices are $n\alpha$=1.665, $n\beta$=1.672 - 1.676, and $n\gamma$=1.684 - 1.686. It is typically almost colourless in thin section, but some lawsonite is pleochroic from colourless to pale yellow to pale blue, depending on orientation. The mineral has a Mohs hardness of 8 and a specific gravity of 3.09. It has perfect cleavage in two directions and a brittle fracture.

Lawsonite is a metamorphic mineral typical of the blueschist facies. It also occurs as a secondary mineral in altered gabbro and diorite. Associate minerals include epidote, titanite, glaucophane, garnet and quartz. It is an uncommon constituent of eclogite.

It was first described in 1895 for occurrences in the Tiburon peninsula, Marin County, California. It was named for geologist Andrew Lawson (1861–1952) of the University of California by two of Lawson's graduate students, Charles Palache and Frederick Leslie Ransome.

### *Composition*

Lawsonite is a metamorphic silicate mineral related chemically and structurally to the epidote group of minerals. It is close to the ideal composition of $CaAl_2Si_2O_7(OH)_2 \cdot H_2O$ giving it a close chemical composition with anorthite $CaAl_2Si_2O_8$ (its anhydrous equivalent), yet lawsonite has greater density and a different Al coordination (Comodi et al., 1996). The substantial amount of water bound in lawsonite's crystal structure is released during its breakdown to denser minerals during prograde metamorphism. This means lawsonite is capable of conveying appreciable water to shallow depths in subducting oceanic lithosphere (Clark et al., 2006). Experimentation on lawsonite to vary its responses at different temperatures and different pressures is among its most studied aspects, for it is these qualities that affect its abilities to carry water down to mantle depths, similar to other OH-containing phases like antigorite, talc, phengite, staurolite, and epidote (Comodi et al., 1996).

### *Geologic Occurrence*

Lawsonite is a very widespread mineral and has attracted considerable interest over the last few years because of its importance as a marker of moderate pressure (6-12 kb) and low Temperature (300 - 400 °C) conditions in nature (Clark et al., 2006). This mainly occurs along continental margins (subduction zones) such as those found in: the Franciscan Formation in California at Reed Station, Tiburon Peninsula of Marin County, California; the Piedmont metamorphic rocks of Italy; and schists in New Zealand, New Caledonia, China, Japan and from various points in the circum-Pacific orogenic belt.

### *Crystal Structure*

Though lawsonite and anorthite have similar compositions, their structures are quite different. While anorthite has a tetrahedral coordination with Al (Al substitutes for Si in feldspars), lawsonite has an octahedral coordination with Al, making it an orthorhombic sorosilicate with a space group of Cmcm which consists of $Si_2O_7$ Groups and O, OH, F, and $H_2O$ with cations in [4] and/or > [4] coordination. This is much similar to the epidote group which lawsonite is often found in conjunction with, which are also sorosilicates because their structure consists of two connected SiO4 tetrahedra plus connecting cation. The water contained in its structure is made possible by cavities formed by rings of two Al octahedral and two $Si_2O_7$ groups, each containing an isolated water molecule and calcium atom. The hydroxyl units are bound to the edge-sharing Al octahedral (Baur, 1978).

### *Physical Properties*

Lawsonite has crystal habits of orthorhombic prismatic, which are crystals shaped like slender prisms, or tubular figures, which are form dimensions that are thin in one direction, both with two perfect cleavages. This crystal is transparent to translucent and varies in colour from white to pale blue to colourless with a white streak and a vitreous or greasy luster. It has a relatively low specific gravity of 3.1g/cm3, and a pretty high hardness of 7.5 on Mohs scale of hardness, slightly higher than quartz. Under the microscope, lawsonite can be seen as blue, yellow, or colourless under plane polarized light while the stage is rotated. Lawsonite has three refractive indices of $n\alpha$ = 1.665 $n\beta$ = 1.672 - 1.676 $n\gamma$ = 1.684 - 1.686, which produces a birefringence of $\delta$ = 0.019 - 0.021 and an optically positive biaxial interference figure.

### *Significance of Lawsonite*

Lawsonite is a significant metamorphic mineral as it can be used as an index mineral for high pressure conditions. Index minerals are used in geology to determine the degree of metamorphism a rock has experienced. New metamorphic minerals form through solid-state cation exchanges following changing pressure and temperature conditions imposed upon the protolith (pre-metamorphosed rock). This new mineral that is produced in the metamorphosed rock is the index mineral, which indicates the minimum pressure and temperature the protolith must have achieved in order for that mineral to form.

Lawsonite is known to form in high pressure, low temperature conditions, most commonly found in subduction zones where cold oceanic crust subducts down oceanic trenches into the mantle (Comodi et al., 1996). The initially low temperature of the slab, and fluids taken down with it manage to depress isotherms and keep the slab much colder than the surrounding mantle, allowing for these unusual high pressure, low temperature conditions. Glaucophane, kyanite and zoisite are other common minerals in the blueschist facies and are commonly found to coexist (Pawley et al., 1996). This assemblage is diagnostic of this facies.

## Manganvesuvianite

Manganvesuvianite is a rare mineral with formula $Ca_{19}Mn^{3+}(Al,Mn^{3+},Fe^{3+})_{10}(Mg,Mn^{2+})_2(Si_2O_7)_4(SiO_4)_{10}O(OH)_9$. The mineral is red to nearly black in colour. Discovered in South Africa and described in 2002, it was so named for the prevalence of manganese in its composition and its relation to vesuvianite.

### *Occurrence and Formation*

Manganvesuvianite crystals occur as long prisms up to 1.5 cm (0.59 in). Small crystals are transparent and red to lilac in colour; large crystals are opaque and nearly black in colour with dark-red internal reflections. Strongly zoned crystals less than 0.2 mm (0.0079 in) in size constitute rock-forming manganvesuvianite.

As of 2012, manganvesuvianite has been found at two locations in South Africa. It formed at temperatures of 250 to 400 °C (482 to 752 °F) by the hydrothermal alteration of sedimentary and metamorphic manganese ores. Crystallization occurred in fault planes and lenticular bodies in the ore bed or by filling veins and vugs. Manganvesuvianite has been found in association with calcite, manganese-poor grossular, hydrogrossular-henritermierite, mozartite, serandite-pectolite, strontiopiemontite-tweddillite, and xonotlite.

Manganvesuvianite is a member of the vesuvianite group and is the manganese analogue of vesuvianite.

### History

In 1883, Arnold von Lasaulx made the first detailed description of vesuvianite containing up to 3.2 wt% MnO from Lower Silesia in Poland. Studies in the 1980s and 1990s revealed that the vesuvianite group was more complex than previously assumed, necessitating the definition of new minerals. In 2000, vesuvianite was found containing up to 14.3 wt% MnO from the Kalahari manganese fields of Northern Cape Province, South Africa. Manganvesuvianite proper was discovered in the Wessels (27°62 56.433 S 22°512 27.873 E) and N'Chwaning (shaft II; 27°82 6.843 S 22°512 55.993 E) mines of the Kalahari manganese fields and described in 2002 in the journal *Mineralogical Magazine*. It was named *manganvesuvianite* for the significant manganese in its formula and its relation to vesuvianite. The mineral and name were approved by the IMA Commission on New Minerals and Mineral Names (IMA 2000-40). The type specimen from the N'Chwaning II Mine is held at the Natural History Museum of Bern in Switzerland.

## Melilite

Melilite refers to a mineral of the melilite group. Minerals of the group are solid solutions of several endmembers, the most important of which are gehlenite and åkermanite. A generalized formula for common melilite is $(Ca,Na)_2(Al,Mg,Fe^{2+})[(Al,Si)SiO_7]$. Discovered in 1793 near Rome, it has a yellowish, greenish brown colour. The name derives from the Greek words meli "honey" and lithos "stone".

Minerals of the melilite group are sorosilicates. They have the same basic structure, of general formula $A_2B(T_2O_7)$. The melilite structure consist of pairs of fused $TO_4$, where $T$ may be Si, Al, B, in bow-tie form. Sharing one corner, the formula of the pair is $T_2O_7$. These bow-ties are linked together into sheets by the $B$ cations. The sheets are held together by the $A$ cations, most commonly calcium and sodium. Aluminium may sit on either the $T$ or the $B$ site.

Minerals with the melilite structure may show a cleavage parallel to the (001) crystallographic directions and may show weaker cleavage perpendicular to this, in the {110} directions. Melilite is tetragonal.

The important endmembers of common melilite are åkermanite $Ca_2Mg(Si_2O_7)$ and gehlenite $Ca_2Al[AlSiO_7]$. Many melilites also contain appreciable iron and sodium.

Some other compositions with the melilite structure include: alumoåkermanite $(Ca,Na)_2(Al,Mg,Fe^{2+})(Si_2O_7)$, okayamalite $Ca_2B[BSiO_7]$, gugiaite $Ca_2Be[Si_2O_7]$, hardystonite $Ca_2Zn[Si_2O_7]$, barylite $BaBe_2[Si_2O_7]$, andremeyerite $BaFe_2{}^{+2}[Si_2O_7]$. Some structures formed by replacing one oxygen by F or OH: leucophanite $(Ca,Na)_2(Be,Al)[Si_2O_6(F,OH)]$, jeffreyite $(Ca,Na)_2(Be,Al)[Si_2O_6(O,OH)]$, and meliphanite $(Ca,Na)_2(Be,Al)[Si_2O_6(OH,F)]$.

### *Occurrences*

Melilite with compositions dominated by the endmembers akermanite and gehlenite is widely distributed but uncommon. It occurs in metamorphic and igneous rocks and in meteorites.

Typical metamorphic occurrences are in high-temperature metamorphosed impure limestones. For instance, melilite occurs in some high-temperature skarns.

Melilite also occurs in unusual silica-undersaturated igneous rocks. Some of these rocks appear to have formed by reaction of magmas with limestone. Other igneous rocks containing melilite crystallize from magma derived from the Earth's mantle and apparently uncontaminated by the Earth's crust. The presence of melilite is an essential constituent in some rare igneous rocks, such as olivine melilitite. Extremely rare igneous rocks contain as much as 70% melilite, together with minerals such as pyroxene and perovskite.

Melilite is a constituent of some calcium- and aluminium-rich inclusions (CAIs) in chondritic meteorites. Isotope ratios of magnesium and some other elements in these inclusions are of great importance in deducing processes that formed our solar system.

## Nabalamprophyllite

Nabalamprophyllite has a general formula of $Ba(Na,Ba)\{Na_3Ti[Ti_2O_2Si_4O_{14}](OH,F)_2\}$ The name is given for its composition (Naba, meaning sodium, Na and barium, Ba) and relation to other lamprophyllite-group minerals. Lamprophyllite is a rare Ti-bearing silicate mineral usually found in intrusive igneous rocks.

Nabalamprophyllite is monoclinic, which means crystallographically, it contains three axes of unequal length and the angles between two of the axes are 90°, and one is less than 90°. It belongs to the space group P2/m. The mineral also has an orthorhombic polytype (nabalamprophyllite-2O) This mineral belongs to the space group Pnmn. In terms of its optical properties, nabalamprophyllite is anisotropic which means the velocity of light varies depending on direction through the mineral. Its calculated relief is 1.86 - 1.87. Its colour in plane polarized light is green-brown, and it is weakly pleochroic.

The mineral has only been found in Russia, usually in association with coarse-grained igneous rocks called pegmatites. The type localities are the Inagli alkaline–ultrabasic massif, Yakutia and the Kovdor alkaline–ultrabasic massif in the Kola Peninsula.

### Normandite

Normandite is a brittle orange brown sorosilicate mineral discovered in 1997 by Charles Normand (born 1963), of Montreal. Normandite occurs in Khibiny Massif, Kola, Russia; in Poudrette quarry, Mont-Saint-Hilaire, Quebec and Tenerife, Canary Islands. It is found in nepheline syenite and in miarolitic cavities in nepheline syenite, associated with nepheline, albite, microcline, aegirine, natrolite, catapleiite, kupletskite, eudialyte, cancrinite, villiaumite, rinkite, and donnayite-(Y).

Normandite has a chemical formula of NaCa $(Mn^{2+},Fe^{2+})$ (Ti,Nb,Zr) $Si_2O_7(O,F)_2$. It crystallizes in the monoclinic-prismatic crystal system. It occurs as transparent to translucent orange-brown aggregates of subparallel acicular crystals up to 10 mm in length, and as patches of yellow, fibrous crystals. It has a white to very pale yellow streak and vitreous luster. It is brittle, with distinct {100} and {001} cleavages, and a conchoidal fracture. It has a specific gravity of 3.48 to 3.5, a Mohs hardness of 5 to 6 and refractive index values of $n\alpha=1.743$, $n\beta=1.785$ and $n\gamma=1.810$. It is named after Charles Normand (born 1963), Canadian geologist.

### Ruizite

Ruizite is a sorosilicate mineral with formula $Ca_2Mn_2Si_4O_{11}(OH)_4{\cdot}2H_2O$. It was discovered at the Christmas mine in Christmas, Arizona, and described in 1977. The mineral is named for discoverer Joe Ana Ruiz.

### Description and Occurrence

Ruizite is translucent and orange to red-brown in colour with an apricot yellow streak. The mineral occurs as euhedral prisms up to 1 millimetre (0.039 inches) or as radial clusters of acicular (needle-like) crystals.

Ruizite is common at the Christmas mine. The mineral is known from Arizona, Pennsylvania, and Northern Cape Province, South Africa. Ruizite occurs in association with apophyllite, bornite, calcite, chalcopyrite, datolite, diopside, grossular, inesite, junitoite, kinoite, orientite, pectolite, quartz, smectite, sphalerite, vesuvianite, and wollastonite. Ruizite is found in veinlets or fracture surfaces of limestone metamorphosed into a calc-silicate assemblage. The mineral formed by retrograde metamorphism during cooling of a calc–silicate skarn assemblage in an oxidizing environment.

### Crystal Structure and Chemistry

Ruizite crystallizes in the monoclinic crystal system and twinning is common along the {100} plane between exactly two crystals. Ruizite's formula was originally identified as $CaMn(SiO_3)_2(OH)_2 \cdot 2H_2O$ in 1977. In 1984, Frank C. Hawthorne revised the formula to $Ca_2Mn_2Si_4O_{11}(OH)_4 \cdot 2H_2O$. Ruizite's structure consists of edge-sharing $Mnö_6$ octahedra, connected at corners into sheets and together into a lattice by clusters of $Si_4O_{11}(OH)_2$.

Nitric acid, hydrochloric acid, and potassium hydroxide have little effect on ruizite at low temperatures but readily dissolve the mineral at elevated temperatures.

### History

During the investigation of junitoite at the Christmas mine in Christmas, Arizona, Joe Ana Ruiz and Robert Jenkins discovered an unknown brown mineral. Mine geologist Dave Cook located better specimens, and it was determined to be a new mineral species. The mineral was named *ruizite* in honour of Joe Ruiz as discoverer. Ruizite's properties were analyzed using a sample provided by Joseph Urban, and it was described in the journal *Mineralogical Magazine* in December 1977. The International Mineralogical Association approved the mineral as IMA 1977-077. Type specimens are housed in the University of Arizona, Harvard University, the National Museum of Natural History, and The Natural History Museum.

## Wadsleyite

Wadsleyite is a high-pressure polymorph of olivine, and is an orthorhombic mineral found in the Peace River meteorite in Alberta, Canada. In phase transformations with increasing pressure from $Mg_2SiO_4$-$Fe_2SiO_4$ (forsterite – fayalite), olivine is transformed to wadsleyite ($\beta$-$Mg_2SiO_4$) and then to a spinel-structured ringwoodite

($\gamma$-$Mg_2SiO_4$). This series of transformations is thought to occur during an extraterrestrial shock event in the meteorite prior to its fall on Earth. With a formula of $(Mg,Fe^{2+})_2(SiO_4)$, its cell parametres are as follows: a = 5.7 Å, b = 11.7 Å and c = 8.24 Å. It is polymorphous with ringwoodite and is found to be stable in the transition zone of the Earth's upper mantle. These regions are from 400–525 kilometres (250–326 mi) in depth. Because of oxygens not bound to silicon in the $Si_2O_7$ groups of wadsleyite, it leaves some oxygen atoms underbonded, and as a result, these oxygens are hydrated easily. As a result, there can be high concentrations of hydrogen atoms in the mineral. Hydrous wadsleyite is a considered a potential site for water storage in the Earth's mantle due to the low electrostatic potential of the underbonded oxygen atoms. Although wadsleyite does not contain H in its chemical formula, it may contain more that 3 percent by weight $H_2O$, and may coexist with a hydrous melt at transition zone pressure-temperature conditions. The water solubility and density of wadsleyite are ultimately affected by the temperature and pressure inside of the Earth.

Wadsleyite was first identified by Ringwood and Major in 1966 and was confirmed to be a stable phase by Akimoto and Sato in 1968.(Horiuchi and Sawamoto, 1981) The phase was originally known as $\beta$-$Mg_2SiO_4$ or "beta-phase " and is a polymorph of olivine, along with minerals ringwoodite. Wadsleyite was named for mineralogist Arthur David Wadsley (1918-1969).

## *Composition*

Wadsleyite is a polymorph of forsterite $Mg_2SiO_4$, an end-member of the solid-solution series of olivine(Horiuchi and Sawamoto, 1981). In the phase transformations of forsterite to fayalite $Mg_2SiO_4$-$Fe_2SiO_4$, this magnesium-rich olivine $\alpha$-$Mg_2SiO_4$ changes to wadsleyite $\beta$-$Mg_2SiO_4$ under certain pressure and temperature conditions, and then with increasing pressure, it transforms to ringwoodite $\gamma$-$Mg_2SiO_4$ which is a spinel structure (Price, Putnis, Agrell and Smith, 1983). Wadsleyite is synthesized stably at 1000–1500 °C (1800–2700 °F) and 13 to 18 GPa of pressure between depths of 410–525 kilometres (250–326 mi). figure of the reference shows the high pressure phases of olivine polymorphs beginning with wadsleyite $\beta$-$Mg_2SiO_4$. Geologically speaking, it is a very fine-grained "reactive" forsterite that had been synthesized from hydrous starting materials.

In values of weight percent oxide, the pure magnesian variety of wadsleyite would be 44.5% $SiO_2$, 52.2% MgO, and 3.33% $H_2O$. The average microprobe analysis for wadsleyite yielded: MgO 38.21, $SiO_2$

38.7, CaO 0.07, $Cr_2O_3$ 0.01, MnO 0.43, GeO 22.37, NiO 0.11 and ZnO 0.10.(Smyth, 1987) A recalculation of the number of cations on the basis of four oxygens will yield MgO 1.51, $SiO_2$ 1.03, CaO 0.0019, $Cr_2O_3$ 0.0002, MnO 0.0096, GeO 0.4032, NiO 0.0023 and ZnO 0.0019. An analysis of trace elements in wadsleyite suggests that there are a number of elements included in it. Results demonstrate traces of rubidium Rb, strontium Sr, barium Ba, titanium Ti, zirconium Zr, niobium Nb, hafnium Hf, tantalum Ta, thorium Th, and uranium U in wadsleyite relative to olivine.(Mibe, Nakai, Orihashi, and Fujii 2006) This information suggests that the concentrations of these elements could be larger than what has been supposed in the transition zone of Earth's upper mantle. Moreover, these results help in understanding chemical differentiation and magmatism inside the Earth (Mibe, Nakai, Orihashi, and Fujii 2006).

Although nominally anhydrous, wadsleyite can incorporate more than 3 percent by weight $H_2O$, which means that it is capable of incorporating more water than Earth's oceans and may be a significant reservoir for H (or water) in the Earth's interior.

## Geologic Occurrence

Wadsleyite was found in Peace River meteorite in Peace River, Alberta, Canada. This meteorite, an L6 hypersthene-olivine chrondite, is believed to have formed at high pressure during an extraterrestrial shock event. It occurs as microcrystalline rock fragments, often not surpassing 0.5 millimetres (0.020 in) in diametre, that pseudomorph pre-existing olivine parts within the mineral.(Price, Putnis, Agrell and Smith, 1983) The meteor or asteroid that impacted the earth generated the mineral phase transformations observed in shocked chrondites of the Peace River meteorite.(Beck, Gillet, Jahn, McMillan, Reynard, Van De Moortele and Wilson, 2007) It contains sulfide-rich veins of olivine and is believed, like other meteorite specimens, to have undergone a shock event, causing the grain components of olivine to transform into significant amounts of high-density wadsleyite.(Price, Putnis, Agrell and Smith, 1983)

### *Structure*

Wadsleyite is a spinelloid, and the structure is based on a distorted cubic-closest packing of oxygen atoms as are the spinels. The a-axis and the b-axis is the half diagonal of the spinel unit. The magnesium and the silicon are completely ordered in the structure. There are three distinct octahedral sites, M1, M2, and M3, and a single tetrahedral site. Wadsleyite is a sorosilicate in which $Si_2O_7$ groups are present (Ashbrook, Berry, Farnanf, Le Polle, Pickard and Wimperise, 2006). There are four distinct oxygen atoms in the structure. O2 is a bridging oxygen shared between two tetrahedra, and O1 is a non-silicate oxygen (not bonded to Si). The potentially hydrated O1 atom lies at the centre of four edge-sharing $Mg^{2+}$ octahedra (Smyth, 1987, 1994). If this oxygen is hydrated (protonated), a Mg vacancy can occur at M3. A structure of $\beta$-$Mg_2SiO_4$ is shown in reference. If water incorporation exceeds about 1.5% the M3 vacancies can order in violation of space group *Imma*, reducing the symmetry to monoclinic *I2/m* with beta angle up to 90.4°.

Wadsleyite II is a separate spinelloid phase that might occur between the fields of wadsleyite and ringwoodite. It has both a single ($SiO_4$) and double ($Si_2O_7$) tetrahedral units. It is a hydrous magnesium-iron silicate with variable composition that occurs between the stability regions of wadsleyite and ringwoodite $\gamma$-$Mg_2SiO_4$.(Kleppe, 2006) One-fifth of the silicon atom is in isolated tetrahedral and four-fifths is in $Si_2O_7$ groups so that the structure can be thought of as a mixture of one-fifth spinel and four-fifths wadsleyite.(Horiuchi and Sawamoto, 1981). In the phase of wadsleyite II, there is considered to be possible host hydrogen in the transition zone of the Earth's mantle. Since forsterite is thought to be about or little over 50% of the mantle, the transition region in the upper mantle could be an important water reservoir.(Kleppe, 2006) Wadsleyite is very water soluble and can accept up to 3 wt. % $H_2O$ as hydroxyl at this site. Its water content is very significant in understanding the way Earth developed. Synthetic hydrous wadsleyite II is pictured in reference. Wadsleyite II in a variably hydrous magnesium-iron silicate phase. It is a potential host for hydrogen in the transition zone of the Earth's mantle. However, if the water composition of wadsleyite surpasses a 0.1–0.2 wt% amount, it could cause partial melting. As a result, an upwelling flow of water could affect the distribution of particular elements in the Earth.(Huang, Karato and Xu, 2005)

## *Crystallography and Physical Properties*

Wadsleyite crystallizes in the orthorhombic crystal system and has a unit cell volume of 550.00 Å³. Its space group is *Imma* and its cell parametres are as follows: a = 5.6921 Å, b = 11.46 Å and c = 8.253 Å.(Price, Putnis, Agrell and Smith, 1983) A more recent structure of wadsleyite confirms the cell parametres to be a = 5.698 Å, b = 11.438 Å and c = 8.257 Å.(Horiuchi and Sawamoto, 1981). Pure magnesian wadsleyite is colourless, but iron bearing varieties are dark green.

The wadsleyite minerals generally have a microcrystalline texture and are fractured. Because of small crystal size, detailed optical data could not be obtained; however, wadsleyite is anisotropic with low first-order birefringence colours.(Price, Putnis, Agrell and Smith, 1983) It is biaxial with a mean refractive index of n = 1.76. I has a calculated specific gravity of 3.84. In X-ray powder diffraction, its strongest points in pattern are: 2.886(50)(040), 2.691(40)(013), 2.452(100,141), 2.038(80)(240), 1.442(80)(244).(Price, Putnis, Agrell and Smith, 1983)

### *Biographic Sketch*

Arthur David Wadsley (1918-1969) received the privilege of getting a mineral named after him due his contributions to geology such as the crystallography of minerals and other inorganic compounds.(Price, Putnis, Agrell and Smith, 1983) The proposal to have wadsleyite named after Wadsley was approved by the Commission on New Minerals and Mineral Names of the International Mineralogical Association.(Price, Putnis, Agrell and Smith, 1983) The type specimen is now preserved in the collection of the Department of Geology at the University of Alberta.

# 7

# Cyclosilicates

## Alluaivite

Alluaivite is an exceedingly rare mineral of the eudialyte group, with complex formula written as $Na_{19}(Ca,Mn)_6(Ti,Nb)_3Si_{26}O_{74}Cl \cdot 2H_2O$. It is unique among the eudialyte group as the only titanosilicate (other representatives of the group are usually zirconosilicates). It is named after Mt. Alluaiv in Lovozero Tundry massif, Kola Peninsula, Russia, where it is found in ultra-agpaitic, hyperalkaline pegmatites.

### *Pabstite*

Pabstite is a barium tin titanium silicate mineral that is found in contact metamorphosed limestone. It belongs to the benitoite group of minerals. The chemical formula of pabstite is $Ba(Sn,Ti)Si_3O_9$. It is found in Santa Cruz, California. The crystal system of the mineral is hexagonal.

### *Composition*

Pabstite is 37.7% $SiO_2$, 3.8% $TiO_2$, 24.4% $SnO_2$ and 33.2% BaO. However, Ti and Sn could vary from point to point by approximately ±0.5% $TiO_2$ and ±1% $SnO_2$. Pabstite is a tin bearing analog of benitoite. Although, ($Sn^{4+}$ = 0.71 Å) and ($Ti^{4+}$= 0.68Å) have similar charge and ionic size, it is uncommon to find them substituting each other.

### *Geologic Occurrence*

Pabstite commonly occurs as anhedral crystals and masses that vary in their colour from colourless to white. They produce a pink tinge when they are freshly broken. Large amounts of pabstite were found in Santa Cruz as fracture filling and dissemnted grains in recrystallized siliceous limestones. This is geologic evidence of contact metamorphism. In addition, pabstite can be found in Rush Creek in California when benitoite contains small amounts of tin. It is commonly occurs in rocks that contain calcite, quartz, tremolite, witherite, phlogopite, diopside, minor amounts of forsterite and taramellite. Pabstite can also be found associated with galena, cassiterite and sphalerite.

### Structure

Pabstite is considered the tin analog of benitoite. It has a hexagonal crystal system with a P6*2C space group. Its dimensions are as follows, a= 6.7037(7), c= 9.824(1) Å3, C= 382.3(1) Å3, Z=2. Pabstite has the structure of the benitoite group of minerals. In the structure of pabstite, there are four oxygens surrounding the cations in a pseudo-tetrahedral arrangement. A three-membered cyclosilicate ring ($Si_3O_9$) is formed by repeating the tetrahedron using space group symmetry. Quadrivalent cations connect the rings to form a three-dimensional framework. Since the silicate rings have a geometry that is identical in all directions, a solid rigid unit is formed in the structure. In a distorted hexagonal antiprism arrangement, Ba, which has a high symmetry, is bounded by 12 oxygens. Two different BaÜÜÜO distances are present. [$M^{4+}$ $Si^3O^9$] frameworks are connected by Ba cations which lead to a poor cleavage.

### Physical Properties

Finding pabstite in the field is hard and rare. When using a shortwave ultraviolet light, bluish white fluorescence is revealed from the specimens. This property is commonly used to identify pabstite. Pabstite is colourless to white. The diametre of pabstite grains is usually less than 2 mm and they contain minute fluid and solid inclusions. Its hardness is 6 on Mohs scale of mineral hardness. Its density is 4.03 g/cm$^3$ and it is uniaxial. The interference colours of pabstite are anomalous blue-violet and golden yellow. The refractive indices of pabstite are $\omega$ = 1.685±0.002 and $\varepsilon$ = 1.674±0.002 which result in low birefringence and absent dichroism.

### Discovery and Locations

Pabstite was first described in 1965 for an occurrence in the Kalkar quarry of Santa Cruz County, California. The mineral was named for Adolf Pabst (1899–1990) a mineralogy professor at the University of California, Berkeley.

Pabstite has also been reported from Tres Pozos, Baja California Norte, Mexico and the Alai Range of the Tien Shan Mountains in Tajikistan.

## Pezzottaite

Pezzottaite, marketed under the name raspberyl or raspberry beryl, is a newly identified mineral species, first recognized by the International Mineralogical Association in September 2003. Pezzottaite is a caesium analogue of beryl, a silicate of caesium, beryllium, lithium and aluminium, with the chemical formula $Cs(Be_2Li)Al_2Si_6O_{18}$.

Named after Italian geologist and mineralogist Federico Pezzotta, pezzottaite was first thought to be either red beryl or a new variety of beryl ("caesium beryl"); unlike other beryls, however, pezzottaite contains lithium and crystallizes in the trigonal crystal system rather than the hexagonal system.

Colours include shades of raspberry red to orange-red and pink. Recovered from miarolitic cavities in the granitic pegmatite fields of Fianarantsoa province, southern Madagascar, the pezzottaite crystals were small—no more than about 7 cm in their widest dimension—and tabular or equant in habit, and few in number, most being heavily included with growth tubes and liquid feathers.

Approximately 10 per cent of the rough material would also exhibit chatoyancy when polished. Most cut pezzottaite gems are under one carat (200 mg) in weight and rarely exceed two carats (400 mg).

With the exception of hardness (8 on Mohs scale), the physical and optical properties of pezzottaite—i.e., specific gravity 3.10 (average), refractive index 1.601 to 1.620, birefringence 0.008 to 0.011 (uniaxial negative)—are all higher than typical beryl. Pezzottiate is brittle with a conchoidal to irregular fracture, and streaks white. Like beryl, it has an imperfect to fair basal cleavage. Pleochroism is moderate, from pink-orange or purplish pink to pinkish purple. Pezzottaite's absorption spectrum, as seen by a hand-held (direct vision) spectroscope, features a band at 485–500 nm with some specimens showing additional weak lines at 465 and 477 nm and a weak band at 550 to 580 nm.

Most (if not all) of the Madagascan deposits have since been exhausted. Pezzottaite has been found in at least one other locality, Afghanistan: this material was first thought to be caesium-rich morganite (pink beryl). Like morganite and bixbite, pezzottaite is believed to owe its colour to radiation-induced colour centres involving trivalent manganese. Pezzottaite will lose its colour if heated to 450 °C for two hours, but the colour can be restored with gamma irradiation.

### *Sekaninaite*

Sekaninaite ($(Fe^{+2},Mg)_2Al_4Si_5O_{18}$) is a silicate mineral, the iron rich analogue of cordierite.

It was first described in 1968 for an occurrence in Dolní Bory, Vyso0ina Region, Moravia, Czech Republic, and is now known also from Ireland, Japan, and Sweden. It was named after a Czech mineralogist, Josef Sekanina (1901–1986). In Brockley, Ireland sekaninaite occurs in bauxitic clay within the contact aureole of a diabase intrusive plug.

### *Structure and Composition*

***The Chemical Formula of Sekaninaite is:*** (Fe2+, Mg2+)2Al4Si5O18*nH2O. Grapes calculated the percentage weights of the sample from Dolni Bory, This compound exists in nature in the form of two polymorphs: one having a disordered hexagonal structure and the other arranged in an ordered orthorhombic structure. As an aluminosilicate, the repeated and ordered structure is based on polymerization of one or the other's tetrahedral framework of Si, Al tetrahedra (Yakubovich, 2003). Nearly all analyses show excess of Al and deficiency in Si with respect to tetrahedral components. The overall substitution of alkalis causes excess in cations found in (K2O, Na2O, CaO), implying that sekaninaite is essentially anhydrous (Grapes, 2010).

The atomic structures of cordierites are interpreted as a continuous series of structures that vary based on the content of octahedrally coordinated Mg and Fe cations. The varying content of atoms in the octahedral M position has an effect on the orthorhombic unit cell's parametres. The wide range of isomorphism of Mg and Fe(4-96%) suggest the existence of a continuous isomorphic series cordierite (Mg,Fe)2[Al4Si4O18]*nH2O-sekaninaite (Fe,Mg)2[Al4Si4O18]*nH2O. It is shown via crystallographic data that a shift in the iron content leads to a corresponding variance in a and b unit cell parametres (Yakubovich, 2003). As an aluminosilicate/cyclosilicate, the octahedral M-O distances consist of 5 independent tetrahedra form a 3-dimensional anionic framework of ordered and distributed Al3+ and Si4+ cations. One independent AlO4 and two SiO¬4 vortex-sharing tetrahedra share oxygen atoms to form six-member rings along the c axis of the unit cell. Mg, Fe octahedra share edges with SiO4 to form rings from alternating octahedra and tetrahedra. Thus, the framework can be described as a semi-layered structure formed of layers of tetrahedra linked into rings by sharing vertices and octahedra and tetrahedra sharing edges, alternating along the c axis. The distortion of the orthorhombic unit cell is determined by the chemical composition rather than the degree of ordering in the tetrahedral framework (Yakubovich, 2003). The temperature at which the liquidous phases crystallize in a sequence: mullite + tridymite, followed by sekaninaite and finally fayalite + clinoferrosilite (Grapes, 2010). Similar trends are observed for amphiboles, clinopyroxenes, olivines', and others. The increase in the Fe mole fraction of minerals was not related with iron input, but was caused by its redistribution during contact metamorphism (Korchak, 2010).

### *Physical Properties*

Stanek and Miskovsky (1975) first identified and diagnosed sekaninaite as a new mineral in the cordierite series. They sampled the poorly developed crystals of the Dolni Bory region, Czechoslovakia, where specimen did not exceed 70 cm. Dolni Bory samples are very different than samples found in the Kuznetsk paralavas. They are very close analogues with respect to Mg/Fe ratios but vastly different a-, b- and c- parametres (Grapes, 2010). Grapes and colleagues calculated cell dimension to be a 17.230(5), b 9.835(3), c 9.314(3) A. The colour of sekaninaite is bright blue and distinctly pleochroic with X = colourless; Y = blue; Z = pale blue; absorption occurs in the sequence Y > Z > X. Sekaninaite has a hardness of 7 7.5; it cleaves imperfectly along {100} and exhibits parting on {001} (Fleischer, 1977). Majority of crystals show zonation (Fe increasing from core to rim). It common twinned on {110} and {310}, simulating hexagonal symmetry. Sekaninaite is classified under the space group Cccm; it is an orthorhombic crystal that is found in series with cordierite (Stanek, 1975).

## Geologic Occurrence and Location

Sekaninaite was first discovered in the Dolni Bory region of the Czech Republic. Its occurrence is in the albite zone of pegmatite in granulites and gneisses (Fleischer, 1977). Sekaninaite is found in pyrometamorphic rocks, extensively rocks formed via process of ancient combustion metamorphism; paralavas, clinkers and buchites. These combustion metamorphic rocks occur in clinker beds and breccias of vitrified sandstone-siltstone clinker fragments cemented by paralava. These partially baked and oxidized psammitic-pelitic sediments are associated with burnt coal seams, belonging to places like the Kuznetsk coal basin, Siberia (Grapes, 2010). Sekaninaite-Fe-cordierite exists in series and is largely dependent upon variations in solid solution. These minerals are more prevalent in paralavas found in: Power River, Wyoming, Ravat area, Tajikistan, Kenderlyk Basin, eastern Kazakhstan and the Djhar basin in India; each differ in sedimentary mineral assemblage and results depend on high-temperature fusion of mixtures of sandstone-siltstone and minor ferruginous components (Grapes, 2010). These Fe-rich paralavas are composed of Fe-olivine, esseneite, dorite, melilite, Fe-cordierite, anorthite, spinel, tridymite, fayalite, magnetite, quartz etc. (Novikova, 2008).

## Beryl

In geology, beryl is a mineral composed of beryllium aluminium cyclosilicate with the chemical formula $Be_3Al_2(SiO_3)_6$. The hexagonal crystals of beryl may be very small or range to several metres in size. Terminated crystals are relatively rare. Pure beryl is colourless, but it is frequently tinted by impurities; possible colours are green, blue, yellow, red, and white.

### *Etymology*

The name beryl is derived (via Latin: *beryllus*, Old French: *beryl*, and Middle English: beril) from Greek *beryllos* which referred to a "precious blue-green colour-of-sea-water stone" and originated from Prakrit velruliya and Pali veluriya; veliru; from Sanskrit vaidurya-, which is ultimately of Dravidian origin, maybe from the name of Belur or "Velur" in southern India. The term was later adopted for the mineral beryl more exclusively. The Late Latin word *berillus* was abbreviated as *brill-* which produced the Italian word *brillare* meaning "shine", the French word *brille* meaning "shine", the Spanish word *brillo*, also meaning "shine", and the English word *brilliance.*

### *Deposits*

Beryl of various colours is found most commonly in granitic pegmatites, but also occurs in mica schists in the Ural Mountains, and limestone in Colombia. Beryl is often associated with tin and tungsten ore bodies. Beryl is found in Europe in Norway, Austria, Germany, Sweden (especially morganite), Ireland and Russia, as well as Brazil, Colombia, Madagascar, Mozambique, South Africa, the United States, and Zambia. US beryl locations are in California, Colourado, Connecticut, Idaho, Maine, New Hampshire, North Carolina, South Dakota and Utah.

New England's pegmatites have produced some of the largest beryls found, including one massive crystal from the Bumpus Quarry in Albany, Maine with dimensions 5.5 by 1.2 m (18 by 3.9 ft) with a mass of around 18 metric tons; it is New Hampshire's state mineral. As of 1999, the world's largest known naturally occurring crystal of any mineral is a crystal of beryl from Malakialina, Madagascar, 18 metres long and 3.5 metres in diametre, and weighing 380,000 kilograms.

### Varieties

***Aquamarine and Maxixe:*** Aquamarine (from Latin: *aqua marina*, "water of the sea") is a blue or turquoise variety of beryl. It occurs at most localities which yield ordinary beryl. The gem-gravel placer deposits of Sri Lanka contain aquamarine. Clear yellow beryl, such as that occurring in Brazil, is sometimes called *aquamarine chrysolite*. The deep blue version of aquamarine is called *maxixe*. Maxixe is commonly found in the country of Madagascar. Its colour fades to white when exposed to sunlight or is subjected to heat treatment, though the colour returns with irradiation.

The pale blue colour of aquamarine is attributed to $Fe^{2+}$. The $Fe^{3+}$ ions produce golden-yellow colour, and when both $Fe^{2+}$ and $Fe^{3+}$ are present, the colour is a darker blue as in maxixe. Decolouration of maxixe by light or heat thus may be due to the charge transfer $Fe^{3+}$ and $Fe^{2+}$. Dark-blue maxixe colour can be produced in green, pink or yellow beryl by irradiating it with high-energy particles (gamma rays, neutrons or even X-rays).

In the United States, aquamarines can be found at the summit of Mt. Antero in the Sawatch Range in central Colorado. In Wyoming, aquamarine has been discovered in the Big Horn Mountains, near Powder River Pass. In Brazil, there are mines in the states of Minas Gerais, Espírito Santo, and Bahia, and minorly in Rio Grande do Norte. The mines of Colombia, Zambia, Madagascar, Malawi, Tanzania and Kenya also produce aquamarine.

The largest aquamarine of gemstone quality ever mined was found in Marambaia, Minas Gerais, Brazil, in 1910. It weighed over 110 kg, and its dimensions were 48.5 cm (19 in) long and 42 cm (17 in) in diametre. The largest cut aquamarine gem is the Dom Pedro aquamarine, now housed in the Smithsonian Institution's National Museum of Natural History.

### Red Beryl

Red beryl (also known as "red emerald" or "scarlet emerald") is a red variety of beryl. It was first described in 1904 for an occurrence, its type locality, at Maynard's Claim (Pismire Knolls), Thomas Range, Juab County, Utah. The old synonym "bixbite" is deprecated from the CIBJO, because of the risk of confusion with the mineral bixbyite (also named after the mineralogist Maynard Bixby). The dark red colour is attributed to $Mn^{3+}$ ions.

***Figure:*** *Red beryl*

Red beryl is very rare and has only been reported from a handful of locations including: Wah Wah Mountains, Beaver County, Utah; Paramount Canyon and Round Mountain, Sierra County, New Mexico; and Juab County, Utah. The greatest concentration of gem-grade red beryl comes from the Violet Claim in the Wah Wah Mountains of mid-western Utah, discovered in 1958 by Lamar Hodges, of Fillmore, Utah, while he was prospecting for uranium. Prices for top quality natural red beryl can be as high as $10,000 per carat for faceted stones. Red beryl has been known to be confused with pezzottaite, also known as raspberry beryl or "raspberyl", a gemstone that has been found in Madagascar and now Afghanistan – although cut gems of the two varieties can be distinguished from their difference in refractive index.

While gem beryls are ordinarily found in pegmatites and certain metamorphic stones, red beryl occurs in topaz-bearing rhyolites. It is formed by crystallizing under low pressure and high temperature from a pneumatolitic phase along fractures or within near-surface miarolitic cavities of the rhyolite. Associated minerals include bixbyite, quartz, orthoclase, topaz, spessartine, pseudobrookite and hematite.

## Cordierite

Cordierite (mineralogy) or iolite (gemology) is a magnesium iron aluminium cyclosilicate. Iron is almost always present and a solid solution exists between Mg-rich cordierite and Fe-rich sekaninaite with a series formula: $(Mg,Fe)_2Al_3(Si_5AlO_{18})$ to $(Fe,Mg)_2Al_3(Si_5AlO_{18})$. A high temperature polymorph exists, indialite, which is isostructural with beryl and has a random distribution of Al in the $(Si,Al)_6O_{18}$ rings.

### *Name and Discovery*

Cordierite, which was discovered in 1813, is named after the French geologist Louis Cordier (1777–1861).

### *Occurrence*

Cordierite typically occurs in contact or regional metamorphism of argillaceous rocks. It is especially common in hornfels produced by contact metamorphism of pelitic rocks. Two common metamorphic mineral assemblages include sillimanite-cordierite-spinel and cordierite-spinel-plagioclase-orthopyroxene. Other associated minerals include garnet (cordierite-garnet-sillimanite gneisses) and anthophyllite. Cordierite also occurs in some granites, pegmatites, and norites in gabbroic magmas. Alteration products include mica, chlorite, and talc. Cordierite occurs in the granite contact zone at Geevor Tin Mine in Cornwall.

### *Commercial Use*

Catalytic converters are commonly made from ceramics containing a large proportion of synthetic cordierite. The manufacturing process deliberately aligns the cordierite crystals to make use of the very low thermal expansion seen for one axis. This prevents thermal shock cracking from taking place when the catalytic converter is used.

### *Gem Variety*

As the transparent variety iolite, it is often used as a gemstone. The name "iolite" comes from the Greek word for violet. Another old name is *dichroite*, a Greek word meaning "two-coloured rock", a reference to cordierite's strong pleochroism. It has also been called "water-sapphire" and "Vikings' Compass" because of its usefulness in determining the direction of the sun on overcast days, the Vikings having used it for this purpose. This works by determining the direction of polarization of the sky overhead. Light scattered by air molecules is polarized, and the direction of the polarization is at right angles to a line to the sun, even when the sun's disk itself is obscured by dense fog or lies just below the horizon.

Gem quality iolite varies in colour from sapphire blue to blue violet to yellowish gray to light blue as the light angle changes. Iolite is sometimes used as an inexpensive substitute for sapphire. It is much softer than sapphires and is abundantly found in Australia (Northern Territory), Brazil, Burma, Canada (Yellowknife area of the Northwest Territories), India, Madagascar, Namibia, Sri Lanka, Tanzania and the United States (Connecticut). The largest iolite crystal found weighed more than 24,000 carats, and was discovered in Wyoming, US.

## Dioptase

Dioptase is an intense emerald-green to bluish-green copper cyclosilicate mineral. It is transparent to translucent. Its luster is vitreous to sub-adamantine. Its formula is $CuSiO_3 \cdot H_2O$ (also reported as $CuSiO_2(OH)_2$). It has a hardness of 5, the same as tooth enamel. Its specific gravity is 3.28–3.35, and it has two perfect and one very good cleavage directions. Additionally, dioptase is very fragile and specimens must be handled with great care. It is a trigonal mineral, forming 6-sided crystals that are terminated by rhombohedra.

### *History*

Late in the 18th century, copper miners at the Altyn-Tyube (Altyn-Tube) mine, Karagandy Province, Kazakhstan thought they found the emerald deposit of their dreams. They found fantastic cavities in quartz veins in a limestone, filled with thousands of lustrous emerald-green transparent crystals. The crystals were dispatched to Moscow, Russia for analysis. However the mineral's inferior hardness of 5 compared with emerald's greater hardness of 8 easily distinguished it. Later Fr. René Just Haüy (the famed French mineralogist) in 1797 determined that the enigmatic Altyn-Tyube mineral was new to science and named it dioptase (Greek, *dia*, "through" and *optima*, "vision"), alluding to the mineral's two cleavage directions that are visible inside unbroken crystals.

### *Occurrence*

Dioptase is an uncommon mineral found mostly in desert regions where it forms as a secondary mineral in the oxidized zone of copper sulfide mineral deposits. However, the process of its formation is not simple, the oxidation of copper sulfides should be insufficient to crystallize dioptase as silica is normally minutely soluble in water except at highly alkaline pH. The oxidation of sulfides will generate highly acidic fluids rich in sulfuric acid that should suppress silica solubility. However, in dry climates and with enough time, especially in areas of a mineral deposit where acids are buffered by carbonate, minute quantities of silica may react with dissolved copper forming dioptase and chrysocolla.

The Altyn Tube mine in Kazakhstan still provides handsome specimens; a brownish quartzite host distinguishes its specimens from other localities. The finest specimens of all were found at the Tsumeb Mine in Tsumeb, Namibia. Tsumeb dioptase is wonderfully lustrous and transparent, with its crystal often perched on an attractive snow-white carbonate matrix. Dioptase is also found in the deserts of the southwestern USA. A notable occurrence is the old Mammoth-Saint Anthony Mine near Mammoth, Arizona where small crystals that make fine micromount specimens are found. In addition, many small, pale-green coloured crystals of dioptase have come from the Christmas Mine near Hayden, Arizona. Another classic locality for fine specimens is Renéville, Congo-Brazzaville. Finally, an interesting occurrence is the Malpaso Quarry in Argentina. Here tiny bluish-green dioptase is found on and in quartz. It appears at this occurrence, dioptase is primary and has crystallized with quartz, native copper, and malachite.

### *Use*

Dioptase is popular with mineral collectors and it is occasionally cut into small emerald-like gems. Dioptase and chrysocolla are the only relatively common copper silicate minerals. A dioptase gemstone should never be exposed to ultrasonic cleaning or the fragile gem will shatter. As a ground pigment, dioptase can be used in painting.

## Emerald

Emerald is a gemstone, and a variety of the mineral beryl ($Be_3Al_2(SiO_3)_6$) coloured green by trace amounts of chromium and sometimes vanadium. Beryl has a hardness of 7.5–8 on the 10-point Mohs scale of mineral hardness. Most emeralds are highly included, so their toughness (resistance to breakage) is classified as generally poor.

### *Etymology*

The word "Emerald" is derived (via Old French: Esmeraude and Middle English: Emeraude), from Vulgar Latin: Esmaralda/ Esmaraldus, a variant of Latin Smaragdus, which originated in Greek: smaragdos; "green gem".

### *Properties Determining Value*

Emeralds, like all coloured gemstones, are graded using four basic parametres–the four Cs of Connoisseurship: *Colour, Cut, Clarity* and *Crystal*. The last C, *crystal*, is simply a synonym for transparency, or what gemologists call *diaphaneity*. Before the 20th century, jewelers used the term *water*, as in "a gem of the finest water", to express the combination of two qualities: colour and crystal. Normally, in the grading of coloured gemstones, colour is by far the most important criterion. However, in the grading of emeralds, crystal is considered a close second. Both are necessary conditions. A fine emerald must possess not only a pure verdant green hue as described below, but also a high degree of transparency to be considered a top gem.

In the 1960s, the American jewelry industry changed the definition of "emerald" to include the green vanadium-bearing beryl as emerald. As a result, *vanadium emeralds* purchased as emeralds in the United States are not recognized as such in the UK and Europe. In America, the distinction between traditional emeralds and the new vanadium kind is often reflected in the use of terms such as "Colombian Emerald".

### *Colour*

Scientifically speaking, colour is divided into three components: *hue, saturation* and *tone*. Emeralds occur in hues ranging from yellow-green to blue-green, with the primary hue necessarily being green. Yellow and blue are the normal secondary hues found in emeralds. Only gems that are medium to dark in tone are considered emerald; light-toned gems are known instead by the species name *green beryl*. The finest emerald are approximately 75% tone on a scale where 0% tone would be colourless and 100% would be opaque black. In addition, a fine stone should be well saturated; the hue of an emerald should be bright (vivid). Gray is the normal saturation modifier or mask found in emerald; a grayish-green hue is a dull green hue.

Emeralds are green by definition (the name is derived from the Greek word "smaragdus", meaning green). Emeralds are the green variety of beryl, a mineral which comes in many other colours that are sometimes also used as gems, such as blue aquamarine, yellow heliodor, pink morganite and colourless goshenite.

### *Clarity*

Emerald tends to have numerous inclusions and surface breaking fissures. Unlike diamond, where the loupe standard, i.e. 10× magnification, is used to grade clarity, emerald is graded by eye. Thus, if an emerald has no visible inclusions to the eye (assuming normal visual acuity) it is considered flawless. Stones that lack surface breaking fissures are extremely rare and therefore almost all emeralds are treated to enhance the apparent clarity. Eye-clean stones of a vivid primary green hue (as described above) with no more than 15% of any secondary hue or combination (either blue or yellow) of a medium-dark tone command the highest prices. This relative crystal non-uniformity makes emeralds more likely than other gemstones to be cut into cabochons, rather than faceted shapes.

### *Treatments*

Most emeralds are oiled as part of the post-lapidary process, in order to improve their clarity. Cedar oil, having a similar refractive index, is often used in this generally accepted practice. Other liquids, including synthetic oils and polymers with refractive indexes close to that of emerald such as *Opticon* are also used. The U.S. Federal Trade Commission requires the disclosure of this treatment when a treated emerald is sold. The use of oil is traditional and largely accepted by the gem trade. Other treatments, for example the use of green-tinted oil, are not acceptable in the trade. The laboratory community has recently standardized the language for grading the clarity of emeralds. Gems are graded on a four step scale; *none*, *minor*, *moderate* and *highly* enhanced. Note that these categories reflect levels of enhancement, not *clarity*. A gem graded *none* on the enhancement scale may still exhibit visible inclusions. Laboratories tend to apply these criteria differently. Some gem labs consider the mere presence of oil or polymers to constitute enhancement. Others may ignore traces of oil if the presence of the material does not materially improve the look of the gemstone.

Given that the vast majority of all emeralds are treated as described above, and the fact that two stones that appear visually similar may actually be quite far apart in treatment level and therefore in value, a consumer considering a purchase of an expensive emerald is well advised to insist upon a treatment report from a reputable gemological laboratory. All other factors being equal, a high quality emerald with moderate enhancement should cost half the price of an identical stone graded none.

### *Emerald Localities*

Emeralds in antiquity were mined in Egypt, India, and Austria.

Colombia is by far the world's largest producer of emeralds, constituting 50–95% of the world production, with the number depending on the year, source and grade. Emerald production in Colombia has increased drastically in the last decade, increasing by 78% from 2000 to 2010. The three main emerald mining areas in Colombia are Muzo, Coscuez, and Chivor. Rare 'trapiche' emeralds are found in Colombia, distinguished by a six-pointed radial pattern made of ray-like spokes of dark carbon impurities.

Zambia is the world's second biggest producer, with its Kafubu River area deposits (Kagem Mines) about 45 km southwest of Kitwe responsible for 20% of the world's production of gem quality stones in 2004. In the first half of 2011 the Kagem mines produced 3.74 tons of emeralds.

Emeralds are found all over the world in countries such as Afghanistan, Australia, Austria, Brazil, Bulgaria, Cambodia, Canada, China, Egypt, Ethiopia, France, Germany, India, Italy, Kazakhstan, Madagascar, Mozambique, Namibia, Nigeria, Norway, Pakistan, Russia, Somalia, South Africa, Spain, Switzerland, Tanzania, United States, Zambia, and Zimbabwe. In the US, emeralds have been found in Connecticut, Montana, Nevada, North Carolina, and South Carolina. In 1997 emeralds were discovered in the Yukon.

### *Synthetic Emerald*

Both hydrothermal and *flux-growth* synthetics have been produced, and a method has been developed for producing an emerald overgrowth on colourless beryl. The first commercially successful emerald synthesis process was that of Carroll Chatham, likely involving a lithium vanadate flux process, as Chatham's emeralds do not have any water and contain traces of vanadate, molybdenum and vanadium. The other large producer of flux emeralds was Pierre Gilson Sr., whose products have been on the market since 1964. Gilson's emeralds are usually grown on natural colourless beryl seeds, which are coated on both sides. Growth occurs at the rate of 1 mm per month, a typical seven-month growth run producing emerald crystals of 7 mm of thickness. Gilson sold his production laboratory to a Japanese firm in the 1980s, but production has since ceased; so has Chatham's, after the 1989 San Francisco earthquake.

Hydrothermal synthetic emeralds have been attributed to IG Farben, Nacken, Tairus, and others, but the first satisfactory commercial product was that of Johann Lechleitner of Innsbruck, Austria, which appeared on the market in the 1960s. These stones were initially sold under the names "Emerita" and "Symeralds", and they were grown as a thin layer of emerald on top of natural colourless beryl stones. Although not much is known about the original process, it is assumed that Leichleitner emeralds were grown in acid conditions. Later, from 1965 to 1970, the Linde Division of Union Carbide produced completely synthetic emeralds by hydrothermal synthesis. According to their patents, acidic conditions are essential to prevent the chromium (which is used as the colourant) from precipitating. Also, it is important that the silicon-containing nutrient be kept away from the other ingredients to prevent nucleation and confine growth to the seed crystals. Growth occurs by a diffusion-reaction process, assisted by convection. The largest producer of hydrothermal emeralds today is Tairus in Russia, which has succeeded in synthesizing emeralds with chemical composition similar to emeralds in alkaline deposits in Colombia, and whose products are thus known as "Colombian Created Emeralds" or "Tairus Created Emeralds". Luminescence in ultraviolet light is considered a supplementary test when making a natural vs. synthetic determination, as many, but not all, natural emeralds are inert to ultraviolet light. Many synthetics are also UV inert.

Synthetic emeralds are often referred to as "created", as their chemical and gemological composition is the same as their natural counterparts. The U.S. Federal Trade Commission (FTC) has very strict regulations as to what can and what cannot be called "synthetic" stone. The FTC says: "§ 23.23(c) It is unfair or deceptive to use the word "laboratory-grown," "laboratory-created," "[manufacturer name]-created," or "synthetic" with the name of any natural stone to describe any industry product unless such industry product has essentially the same optical, physical, and chemical properties as the stone named."

### *Emerald in Different Cultures, and Emerald Lore*

Emerald is regarded as the traditional birthstone for May, as well as the traditional gemstone for the astrological signs of Taurus, Gemini and sometimes Cancer. One of the quainter anecdotes on emeralds was by the 16th-century historian Brantôme, who referred to the many impressive emeralds the Spanish under Cortez had brought back to Europe from Latin America. On one of Cortez's most notable emeralds he had the text engraved *Inter Natos Mulierum non surrexit mayor* ("Among those born of woman there hath not arisen a greater," Matthew 11:11) which referred to John the Baptist. Brantôme considered engraving such a beautiful and simple product of nature sacrilegious and considered this act the cause for Cortez's loss of an extremely precious pearl (to which he dedicated a work, *A beautiful and incomparable pearl*), and even for the death of King Charles IX of France, who died soon after.

# 8

# Inosilicates

## Actinolite

Actinolite is an amphibole silicate mineral with the chemical formula $Ca_2(Mg,Fe)_5Si_8O_{22}(OH)_2$.

### *Etymology*

The name *actinolite* is derived from the Greek word *aktis*, meaning "beam" or "ray", because of the mineral's fibrous nature. (This word is also the origin of the name of the chemical element actinium.)

### *Mineralogy*

Actinolite is an intermediate member in a solid-solution series between magnesium-rich tremolite, $Ca_2Mg_5Si_8O_{22}(OH)_2$, and iron-rich ferro-actinolite, $Ca_2Fe_5Si_8O_{22}(OH)_2$. Mg and Fe ions can be freely exchanged in the crystal structure. Like tremolite, asbestiform actinolite is regulated as asbestos.

### *Occurrence*

Actinolite is commonly found in metamorphic rocks, such as contact aureoles surrounding cooled intrusive igneous rocks. It also occurs as a product of metamorphism of magnesium-rich limestones.

The old mineral name *uralite* is at times applied to an alteration product of primary pyroxene by a mixture composed largely of actinolite. The metamorphosed gabbro or diabase rock bodies, referred to as epidiorite, contain a considerable amount of this *uralitic* alteration.

Fibrous actinolite is one of the six recognised types of asbestos, the fibres being so small that they can enter the lungs and damage the alveoli. Actinolite asbestos was once mined along Jones Creek at Gundagai, Australia.

## *Gemology*

Some forms of actinolite are used as gemstones. One is nephrite, one of the two types of jade (the other being jadeite, a variety of pyroxene).

Another gem variety is the chatoyant form known as *cat's-eye actinolite.* This stone is translucent to opaque, and green to yellowish green colour. This variety has had the misnomer *jade cat's-eye.* Transparent actinolite is rare and is faceted for gem collectors. Major sources for these forms of actinolite are Taiwan and Canada. Other sources are Madagascar, Tanzania, and the United States.

## *Aegirine*

Aegirine is a member of the clinopyroxene group of inosilicates. Aegirine is the sodium endmember of the aegirine-augite series. Aegirine has the chemical formula $NaFeSi_2O_6$ in which the iron is present as $Fe^{3+}$. In the aegirine-augite series the sodium is variably replaced by calcium with iron(II) and magnesium replacing the iron(III) to balance the charge. Aluminium also substitutes for the iron(III). It is also known as *acmite,* which is a fibrous, green-coloured variety.

Aegirine occurs as dark green monoclinic prismatic crystals. It has a glassy luster and perfect cleavage. The Mohs hardness varies from 5 to 6 and the specific gravity is 3.2 to 3.4.

Commonly occurs in alkalic igneous rocks, nepheline syenites, carbonatites and pegmatites. Also in regionally metamorphosed schists, gneisses, and iron formations; in blueschist facies rocks, and from sodium metasomatism in granulites. It may occur as an authigenic mineral in shales and marls. It occurs in association with potassic feldspar, nepheline, riebeckite, arfvedsonite, aenigmatite, astrophyllite, catapleiite, eudialyte, serandite and apophyllite.

Localities include Mont Saint-Hilaire, Quebec, Canada; Kongsberg, Norway; Narsarssuk, Greenland; Kola Peninsula, Russia; Magnet Cove, Arkansas, USA; Kenya; Scotland and Nigeria.

It was first described in 1835 for an occurrence in Rundemyr, Øvre Eiker, Buskerud, Norway. Aegirine was named after Ægir, the Teutonic god of the sea. A synonym for the mineral is *acmite* in reference to the typical pointed crystals.

## Amphibole

Amphibole is the name of an important group of generally dark-coloured, inosilicate minerals, forming prism or needlelike crystals, composed of double chain SiO

4 tetrahedra, linked at the vertices and generally containing ions of iron and/or magnesium in their structures. Amphiboles can be green, black, colourless, white, yellow, blue, or brown.

### *Mineralogy*

Amphiboles crystallize into two crystal systems, monoclinic and orthorhombic. In chemical composition and general characteristics they are similar to the pyroxenes. The chief differences from pyroxenes are that (i) amphiboles contain essential hydroxyl (OH) or halogen (F, Cl) and (ii) the basic structure is a double chain of tetrahedra (as opposed to the single chain structure of pyroxene). Most apparent, in hand specimens, is that amphiboles form oblique cleavage planes (at around 120 degrees), whereas pyroxenes have cleavage angles of approximately 90 degrees. Amphiboles are also specifically less dense than the corresponding pyroxenes. In optical characteristics, many amphiboles are distinguished by their stronger pleochroism and by the smaller angle of extinction (Z angle c) on the plane of symmetry. Amphiboles are the primary constituent of amphibolites.

### *In Rocks*

Amphiboles are minerals of either igneous or metamorphic origin; in the former case occurring as constituents (hornblende) of igneous rocks, such as granite, diorite, andesite and others. Calcium is sometimes a constituent of naturally occurring amphiboles.(C. Michael Hogan. 2010) Those of metamorphic origin include examples such as those developed in limestones by contact metamorphism (tremolite) and those formed by the alteration of other ferromagnesian minerals (hornblende). Pseudomorphs of amphibole after pyroxene are known as uralite.

### *History and Etymology*

The name amphibole (Greek *amphibolos* meaning 'ambiguous') was used by René Just Haüy to include tremolite, actinolite, tourmaline and hornblende. The group was so named by Haüy in allusion to the protean variety, in composition and appearance, assumed by its minerals. This term has since been applied to the whole group. Numerous sub-species and varieties are distinguished, the more important of which are tabulated below in two series. The formulae of each will be seen to be built on the general double-chain silicate formula $RSi_4O_{11}$.

### Mineral Species

Chemical formulae:

Orthorhombic series

- Anthophyllite $(Mg,Fe)_7Si_8O_{22}(OH)_2$
- Holmquistite $Li_2Mg_3Al_2Si_8O_{22}(OH)_2$

Monoclinic series

- Tremolite $Ca_2Mg_5Si_8O_{22}(OH)_2$
- Actinolite $Ca_2(Mg,Fe)_5Si_8O_{22}(OH)_2$
- Cummingtonite $Fe_2Mg_5Si_8O_{22}(OH)_2$
- Grunerite $Fe_7Si_8O_{22}(OH)_2$
- Hornblende $Ca_2(Mg,Fe,Al)_5(Al,Si)_8O_{22}(OH)_2$
- Glaucophane $Na_2(Mg,Fe)_3Al_2Si_8O_{22}(OH)_2$
- Riebeckite (or Crocidolite) $Na_2Fe^{2+}{}_3Fe^{3+}{}_2Si_8O_{22}(OH)_2$
- Arfvedsonite $Na_3Fe^{2+}{}_4Fe^{3+}Si_8O_{22}(OH)_2$
- Richterite $Na_2Ca(Mg,Fe)_5Si_8O_{22}(OH)_2$
- Pargasite $NaCa_2Mg_3Fe^{2+}Si_6Al_3O_{22}(OH)_2$
- Winchite $(CaNa)Mg_4(Al,Fe^{3+})Si_8O_{22}(OH)_2$

### Descriptions

On account of the wide variations in chemical composition, the different members vary considerably in properties and general appearance.

Anthophyllite occurs as brownish, fibrous or lamellar masses with hornblende in mica-schist at Kongsberg in Norway and some other localities. An aluminous related species is known as gedrite and a deep green Russian variety containing little iron as kupfferite.

Hornblende is an important constituent of many igneous rocks. It is also an important constituent of amphibolites formed by metamorphism of basalt.

Actinolite is an important and common member of the monoclinic series, forming radiating groups of acicular crystals of a bright green or greyish-green colour. It occurs frequently as a constituent of greenschists. The name is a translation of the old German word *Strahlstein* (radiated stone).

Glaucophane, crocidolite, riebeckite and arfvedsonite form a somewhat special group of alkali-amphiboles. The first two are blue fibrous minerals, with glaucophane occurring in blueschists and crocidolite (blue asbestos) in ironstone formations, both resulting from dynamo-metamorphic processes. The latter two are dark green minerals, which occur as original constituents of igneous rocks rich in sodium, such as nepheline-syenite and phonolite.

Pargasite is a rare magnesium-rich amphibole with essential sodium, usually found in ultramafic rocks. For instance, it occurs in uncommon mantle xenoliths, carried up by kimberlite. It is hard, dense, black and usually idiomorphic, with a red-brown pleochroism in petrographic thin section.

## Astrophyllite

Astrophyllite is a very rare, brown to golden-yellow hydrous potassium iron titanium silicate mineral. Belonging to the astrophyllite group, astrophyllite may be classed either as an inosilicate, phyllosilicate, or an intermediate between the two. It forms an isomorphous series with kupletskite, to which it is visually identical and often intimately associated. Astrophyllite is of interest primarily to scientists and collectors.

Heavy, soft and fragile, astrophyllite typically forms as *bladed*, radiating *stellate* aggregates. It is this crystal habit that gives astrophyllite its name, from the Greek words *astron* meaning "star" and *phyllon* meaning "leaf". Its great submetallic gleam and darkness contrast sharply with the light (felsic) matrix the mineral is regularly found within. Astrophyllite is usually opaque to translucent, but may be transparent in thin specimens.

As the crystals themselves possess perfect cleavage, they are typically left *in situ*, the entire aggregate often cut into slabs and polished. Owing to its limited availability and high cost, astrophyllite is seldom seen in an ornamental capacity. It is sometimes used in jewellery where it is fashioned into cabochons.

Found in cavities and fissures in unusual felsic igneous rocks, astrophyllite is associated with feldspar, mica, titanite, zircon, nepheline, and aegirine. Common impurities include magnesium, aluminium, calcium, zirconium, niobium, and tantalum. It was first discovered in 1854 at its type locality; Laven Island, Norway. Kupletskite was not known until 1956, over a hundred years later.

Astrophyllite is found in a few scarce, remote localities: Mont-Saint-Hilaire, Quebec, Canada; Pikes Peak, Colorado, USA; Narsarsuk and Kangerdluarsuk, Greenland; Brevig, Norway; and the Kola Peninsula, Russia.

## Balangeroite

Balangeroite is found in one of the most important chrysotile mines in Europe, the Balangero Serpentinite. It is considered an asbestiform in an assemblage of other mineral phases like chrysotile, magnetite and Fe-Ni alloys. In addition to its fibrous occurrence, balangeroite's association with chrysotile raises concerns about its potential toxicity when its fibres are inhaled.

Balangeroite is classified as belonging to one of the two asbestiform silicate groups, the serpentine group. It is intergrown with the dominant member of the serpentinite, chrysotile, often associated with tremolite (a contaminant of chrysotile), which is classified as part of the amphibole group, the other asbestiform silicate. Massive serpentines are economically important for providing building material. The fibrous nature of chrysotile is particularly valuable for thermal insulation purposes, fireproofing etc. Tremolite contaminated chrysotile shows that the toxicity of the asbestos is due to the presence of tremolite and not the entire mass of the chrysotile. Recent publications by Turci have drawn some conclusions that balangeroite contaminated chrysotile does have some areas of concern and can be attributed to the overall toxicity of the airborne fibres in the Balangero mine. Therefore, it cannot be compared to tremolite or croidolite in level of pathogenicity as the two have been proven by autopsies and biopsies to be present in the bodies of the people exposed to their fibres.

### *Composition*

The chemical formula for balangeroite is $(Mg, Fe^{2+}, Fe^{3+}, Mn^{2+})_{42}Si_{16}O_{54}(OH)_{40}$ and it has been calculated as shown in the diagram below by Compagnoni as follows:

**Table:** *Chemical analysis of balangeroite*

| | |
|---|---|
| $SiO_2$ | 28.37 |
| $TiO_2$ | 0.03 |
| $Al_2O_3$ | 0.27 |
| $Fe_2O_3$ | 8.89 |
| $Cr_2O_3$ | 0.03 |
| FeO | 16.95 |
| MnO | 3.59 |

*Contd...*

| | |
|---|---|
| MgO | 31.81 |
| CaO | 0.13 |
| $H_2O$ | 9.93 |
| Total | 100.00 |

Wet chemical, X-ray fluorescence and electron microprobe analyses were used to deduce the composition of balangeroite. The common intergrowth with chrysotile proved to be valuable in providing better chemical resolution as portrayed in table. The results varied due to submicroscopic intergrowths or zoning. From the wet chemical analysis, there was 9.5% average weight loss after calcination at 1000 °C, due to the presence of water. This was calculated as the difference from 100% of microprobe results, with the assumption that large quantities of material usually contain some impurities, and the possible oxidation of $Fe^{2+}$ under heating. A ratio of $Fe^{2+}/Fe^{3+}$ = 2.12 was obtained and on the basis of the known volume and density, the empirical formula for the unit cell was derived ($Mg_{25.70}\, Fe^{2+}_{7.69}\, Fe^{3+}_{3.63}\, Mn^{2+}_{1.65}\, Al_{0.17}\, Ca_{0.07}\, Cr_{0.01}\, Ti_{0.01}$) total= 38.93 $Si_{15.38}O_{53.66}(OH)_{35.92}$.

## Structure

Balangeroite is based on an octahedral build that consists of channels that are filled by chains of silicate tetrahedra grouped in three and 4 rows running along the fibre axis. Balangeroite is isostructural to gageite. In contrast to chrysotile, however, balangeroite has more metal ions than silicon ions and might be in some cases seen as complex iron oxide containing some type of silicate structure in its framework. The surrounding fluid takes in a large number of the cations which are octahedrally coordinated, which unlike amphiboles, may be easily removed. As a consequence, the Mg and Fe are released forcing the silicate structure to become loosely bound and therefore pass into solution. Further tests have been conducted on Balangeroite's ecopersistence and it showed fairly low eco-persistence at neutral pH. Further studies were conducted by imitating weathering in an experiment to predict if weathered fibres retain the toxic potential present in freshly extracted fibres. The tests proved that balangeroite showed removal of Mg and Si which shows a continuous structural severance which extends far beyond the surface.

## Physical Properties

Balangeroite can develop as loose fibres or compact when in large volumes which can be prismatic. Antigorite flakes are included in relict prismatic balangeroite, while transmission electron microscopy observation shows that fibrous balangeroite is partially replaced by chrysotile. The fibres run a couple of centimetres in the [001].

### *Geologic Occurrences*

The piemonte zone, remnant of the Piemontese Ocean from the Late Jurassic, is home to the majority of the serpentines of the Western Alps. The Balangero mine is located in the Lanzu Ultramafic Massif which is in the inner part of the piemonte zone. The Lanzu Ultramafic Massif is believed to have been involved in the subduction processes that were affiliated with the closure of the Piemontese Ocean in the late Jurassic. The earliest generation of metamorphic veins and in particular type 1 Vein that constitute relict prismatic balangeroite (often includes antigorite flakes) were formed during prograde high pressure metamorphism. Fibrous balangeroite is limited to the serpentine-infested rim of the northern Lanzu Ultramafic Massif, with its abundance in the inactive Balangero asbestos mine, where it was discovered.

Balangeroite was named after the location in which it was discovered. Mine workers at the Balangero mine had first discovered it and named it, based on its overall colour and fibrous nature of other minerals present in the mine, xylotile or metaxite. This new mineral, balangeroite, was tested and found to be completely different from xylotile and metaxite in composition as well as optical properties. Balangeroite was already discovered and a somewhat pure specimen was in the Turin University Mineralogy institute's museum since 1925, inventory no. 14873, labelled as “fibrous serpentine (asbestos)- San Vittore, Balangero”.

## Bustamite

Bustamite is a calcium manganese inosilicate (chain silicate) and a member of the wollastonite group. Magnesium, zinc and iron are common impurities substituting for manganese. It is a polymorph of johannsenite, with bustamite as the high-temperature form of $CaMnSi_2O_6$ and johannsenite as the low temperature form. The inversion takes place at 830 °C, but may be very slow.

Bustamite could be confused with light-coloured rhodonite or pyroxmangite, but both these minerals are biaxial (+) whereas bustamite is biaxial (-).

### *Cell Parametres*

There is considerable variety in the literature about the size and type of the unit cell, the formula to be used, and the value of Z, the number of formula units per unit cell.

Bustamite is a triclinic mineral, which could be described by a primitive unit cell, but the larger A-centred cell is often preferred, in order to facilitate comparison with the similar mineral wollastonite.

The formula for bustamite is $CaMn(SiO_3)_2$ but it is sometimes written $(Ca,Mn)SiO_3$, and changing the formula in this way will change the value of Z. The structure is chains of $SiO_4$ tetraheda with repeat unit of three tetrahedra, unlike the pyroxenes where the repeat unit is two. $Ca^{++}$ and $Mn^{++}$ are positioned between the chains. There are 12 tetrahedra in the A-centred unit cell.

The unit cell, the formula and Z cannot be taken separately; they are interlinked and form a consistent set of values. In this article we adopt the A-centred unit cell (space group A1) with a = 7.736 Å, b = 7.157 Å and c = 13.824 Å, the formula $CaMn(SO_3)_2$ and Z = 6. Deer et al take the formula as $(Mn,Ca,Fe)[SiO_3]$ so their value of Z is doubled to 12. Mindat apparently gives the lattice parametres for a face-centred cell, although they give the space group as P1.

## Type Locality

The type locality was originally taken as Tetela de Jonotla, Puebla, Mexico, and the mineral was named for General Anastasio Bustamante (1780–1853), three times President of Mexico. The material from Puebla, however, was later found to be a mixture of johannsenite and rhodonite, so the type locality is now the Franklin Mine, Franklin, Sussex County, New Jersey, USA.

Both bustamite and johansennite are found at Franklin. Bustamite is moderately common there and occurs in a variety of assemblages, associated with rhodonite and tephroite, calcite and tephroite or glaucochroite and tephroite. Vesuvianite, wollastonite, garnet, diopside, willemite, johannsenite, margarosanite and clinohedrite also may be present.

## Environment

Bustamite typically results from metamorphism of manganese-bearing sediments, with attendant metasomatism. At the (new) type locality, Franklin, the oldest rocks are Precambrian gneisses of mixed sedimentary and volcanic origin. Franklin Marble was deposited within these rocks, along with sediments containing zinc, manganese and iron minerals. These sediments were metamorphosed later in the Precambrian, then the rocks were uplifted from the late Precambrian into the Cambrian and quartzite was deposited on the eroded surface. In Cambrian-Ordovician time the quartzite was in turn overlain by limestone, and the rocks have been subject to uplift and erosion up to the present time.

## Diopside

Diopside is a monoclinic pyroxene mineral with composition $MgCaSi_2O_6$. It forms complete solid solution series with hedenbergite ($FeCaSi_2O_6$) and augite, and partial solid solutions with orthopyroxene and pigeonite. It forms variably coloured, but typically dull green crystals in the monoclinic prismatic class. It has two distinct prismatic cleavages at 87 and 93° typical of the pyroxene series. It has a Mohs hardness of six, a Vickers hardness of 7.7 GPa at a load of 0.98 N, and a specific gravity of 3.25 to 3.55. It is transparent to translucent with indices of refraction of $n_\alpha$=1.663–1.699, $n_\beta$=1.671–1.705, and $n_\gamma$=1.693–1.728. The optic angle is 58° to 63°.

### *Formation*

Diopside is found in ultramafic (kimberlite and peridotite) igneous rocks, and diopside-rich augite is common in mafic rocks, such as olivine basalt and andesite. Diopside is also found in a variety of metamorphic rocks, such as in contact metamorphosed skarns developed from high silica dolomites. It is an important mineral in the Earth's mantle and is common in peridotite xenoliths erupted in kimberlite and alkali basalt.

### *Mineralogy and Occurrence*

Diopside is a precursor of chrysotile (white asbestos) by hydrothermal alteration and magmatic differentiation; it can react with hydrous solutions of magnesium and chlorine to yield chrysotile by heating at 600 °C for three days. Some vermiculite deposits, most notably those in Libby, Montana, are contaminated with chrysotile (as well as other forms of asbestos) that formed from diopside.

At relatively high temperatures, there is a miscibility gap between diopside and pigeonite, and at lower temperatures, between diopside and orthopyroxene. The calcium/(calcium+magnesium+iron) ratio in diopside that formed with one of these other two pyroxenes is particularly sensitive to temperature above 900 °C, and compositions of diopside in peridotite xenoliths have been important in reconstructions of temperatures in the Earth's mantle.

Chrome diopside ($(Ca,Na,Mg,Fe,Cr)_2(Si,Al)_2O_6$) is a common constituent of peridotite xenoliths, and dispersed grains are found near kimberlite pipes, and as such are a prospecting indicator for diamonds. Occurrences are reported in Canada, South Africa, Russia, Brazil, and a wide variety of other locations. In the US, chromian diopside localities are described in the serpentinite belt in northern California, in kimberlite in the Colorado-Wyoming State Line district, in kimberlite in the Iron Mountain district, Wyoming, in lamprophyre at Cedar Mountain in Wyoming, and in numerous anthills and outcrops of the Tertiary Bishop Conglomerate in the Green River Basin of Wyoming. Much chromian diopside from the Green River Basin localities and several of the State Line Kimberlites have been gem in character.

### *As a Gem*

Gemstone quality diopside is found in two forms: the black star diopside and the chrome diopside (which includes chromium, giving it a rich green colour). At 5.5–6.5 on the Mohs scale, chrome diopside is relatively soft to scratch. The Mohs scale of hardness does not measure tensile strength or resistance to fracture.

*Violane* is a manganese-rich variety of diopside, violet to light blue in colour.

### *Etymology and History*

Diopside derives its name from the Greek *dis*, "twice", and *òpsθ*, "face" in reference to the two ways of orienting the vertical prism.

Diopside was first described about 1800.

### *Potential Uses*

Diopside based ceramics and glass-ceramics have potential applications in various technological areas. A diopside based glass-ceramic named 'silceram' was produced by scientists from Imperial College, UK during 1980s from blast furnace slag and other waste products. The as produced glass-ceramic is a potential structural material. Similarly, diopside based ceramics and glass-ceramics have potential applications in the field of biomaterials, nuclear waste immobilization and sealing materials in solid oxide fuel cells.

## Enstatite

Enstatite is the magnesium endmember of the pyroxene silicate mineral series enstatite ($MgSiO_3$) - ferrosilite ($FeSiO_3$). The magnesium rich members of the solid solution series are common rock-forming minerals found in igneous and metamorphic rocks. The intermediate composition, $(Mg,Fe)SiO_3$, has historically been known as hypersthene, although this name has been formally abandoned and replaced by orthopyroxene. When determined petrographically or chemically the composition is given as relative proportions of enstatite (En) and ferrosilite (Fs) (e.g., $En_{80}Fs_{20}$).

## Polymorphs and Varieties

Most natural crystals are orthorhombic (space group P*bca*) although three polymorphs are known. The high temperature, low pressure polymorphs are protoenstatite and protoferrosilite (also orthorhombic, space group P*bcn*) while the low temperature forms, clinoenstatite and clinoferrosilite, are monoclinic (space group $P2_1/c$).

Weathered enstatite with a small amount of iron takes on a submetallic luster and a bronze-like colour. This material is termed bronzite, although it is more correctly called altered enstatite.

Bronzite and hypersthene were known long before enstatite, which was first described by G. A. Kenngott in 1855.

An emerald-green variety of enstatite is called chrome-enstatite and is cut as a gemstone. The green colour is caused by traces of chromium, hence the varietal name. In addition, bronzite is also sometimes used as a gemstone.

### *Identification*

Enstatite and the other orthorhombic pyroxenes are distinguished from those of the monoclinic series by their optical characteristics, such as straight extinction, much weaker double refraction and stronger pleochroism. They also have a prismatic cleavage that is perfect in two directions at 90 degrees. Enstatite is white, gray, greenish, or brown in colour; its hardness is 5–6 on the Mohs scale, and its specific gravity is 3.2–3.3.

### *Occurrence*

Isolated crystals are rare, but orthopyroxene is an essential constituent of various types of igneous rocks and metamorphic rocks. Magnesian orthopyroxene occurs in plutonic rocks such as gabbro (norite) and diorite. It may form small idiomorphic phenocrysts and also groundmass grains in volcanic rocks such as basalt, andesite, and dacite.

Enstatite, close to $En_{90}Fs_{10}$ in composition, is an essential mineral in typical peridotite and pyroxenite of the Earth's mantle. Xenoliths of peridotite are common in kimberlite and in some basalt. Measurements of the calcium, aluminum, and chromium contents of enstatite in these xenoliths have been crucial in reconstructing the depths from which the xenoliths were plucked by the ascending magmas.

Orthopyroxene is an important constituent of some metamorphic rocks such as granulite. Orthopyroxene near pure enstatite in composition occurs in some metamorphosed serpentines. Large crystals, a foot in length and mostly altered to steatite, were found in 1874 in the apatite veins traversing mica-schist and hornblende-schist at the apatite mine of Kjörrestad, near Brevig in southern Norway.

Enstatite is a common mineral in meteorites. Crystals have been found in stony and iron meteorites, including one that fell at Breitenbach in the Ore Mountains, Bohemia. In some meteorites, together with olivine it forms the bulk of the material; it can occur in small spherical masses, or chondrules, with an internal radiated structure.

### *Enstatite in Space*

Enstatite is one of the few silicate minerals that have been observed in crystalline form outside our Solar System, particularly around evolved stars and Planetary Nebulae such as NGC 6302. Enstatite is thought to be one of the early stages for the formation of crystalline silicates in space and many correlations have been noted between the occurrence of the mineral and the structure of the object around which it has been observed.

## Jade

Jade is an ornamental stone. The term *jade* is applied to two different metamorphic rocks that are made up of different silicate minerals:

- Nephrite consists of a microcrystalline interlocking fibrous matrix of the calcium, magnesium-iron rich amphibole mineral series tremolite (calcium-magnesium)-ferroactinolite (calcium-magnesium-iron). The middle member of this series with an intermediate composition is called actinolite (the silky fibrous mineral form is one form of asbestos). The higher the iron content the greener the colour.
- Jadeite is a sodium- and aluminium-rich pyroxene. The gem form of the mineral is a microcrystalline interlocking crystal matrix.

### *Etymology*

The English word *jade* (alternative spelling "jaid") is derived (via French *l'ejade* and Latin *ilia*) from the Spanish term *piedra de ijada* (first recorded in 1565) or "loin stone", from its reputed efficacy in curing ailments of the loins and kidneys. *Nephrite* is derived from *lapis nephriticus*, the Latin version of the Spanish *piedra de ijada.*

### *Overview*

***Nephrite and Jadeite:*** Nephrite and jadeite were used from prehistoric periods for hardstone carving. Jadeite has about the same hardness as quartz, while nephrite is somewhat softer. It was not until the 19th century that a French mineralogist determined that "jade" was in fact two different minerals.

Among the earliest known jade artifacts excavated from prehistoric sites are simple ornaments with bead, button, and tubular shapes. Additionally, jade was used for adze heads, knives, and other weapons, which can be delicately shaped. As metal-working technologies became available, the beauty of jade made it valuable for ornaments and decorative objects. Jadeite measures between 6.0 and 7.0 Mohs hardness, and nephrite between 6.0 and 6.5, so it can be worked with quartz or garnet sand, and polished with bamboo or even ground jade.

### *Unusual Varieties*

Nephrite can be found in a creamy white form (known in China as "mutton fat" jade) as well as in a variety of green colours, whereas jadeite shows more colour variations, including blue, lavender-mauve, pink, and emerald-green colours. Of the two, jadeite is rarer, documented in fewer than 12 places worldwide. Translucent emerald-green jadeite is the most prized variety, both historically and today. As "quetzal" jade, bright green jadeite from Guatemala was treasured by Mesoamerican cultures, and as "kingfisher" jade, vivid green rocks from Burma became the preferred stone of post-1800 Chinese imperial scholars and rulers. Burma (Myanmar) and Guatemala are the principal sources of modern gem jadeite. In the area of Mogaung in the Myitkyina District of Upper Burma, jadeite formed a layer in the dark-green serpentine, and has been quarried and exported for well over a hundred years. Canada provides the major share of modern lapidary nephrite. Nephrite jade was used mostly in pre-1800 China as well as in New Zealand, the Pacific Coast and Atlantic Coasts of North America, Neolithic Europe, and Southeast Asia. In addition to Mesoamerica, jadeite was used by Neolithic Japanese and European cultures.

### *History*

***Prehistoric and Historic China:*** During Neolithic times, the key known sources of nephrite jade in China for utilitarian and ceremonial jade items were the now depleted deposits in the Ningshao area in the Yangtze River Delta (Liangzhu culture 3400–2250 BC) and in an area of the Liaoning province and Inner Mongolia (Hongshan culture 4700–2200 BC). Dushan Jade was being mined as early as 6000 BC. In the Yin Ruins of the Shang Dynasty (1600 to 1050 BC) in Anyang, Dushan Jade ornaments were unearthed in the tomb of the Shang kings. Jade was used to create many utilitarian and ceremonial objects, from indoor decorative items to jade burial suits. Jade was considered the "imperial gem". From the earliest Chinese dynasties to the present, the jade deposits most in use were not only those of Khotan in the Western Chinese province of Xinjiang but other parts of China as well, such as Lantian, Shaanxi. There, white and greenish nephrite jade is found in small quarries and as pebbles and boulders in the rivers flowing from the Kuen-Lun mountain range eastward into the Takla-Makan desert area. The river jade collection is concentrated in the Yarkand, the White Jades (Yurungkash) and Black Jade (Karakash) Rivers. From the Kingdom of Khotan, on the southern leg of the Silk Road, yearly tribute payments consisting of the most precious white jade were made to the Chinese Imperial court and there worked into *objets d'art* by skilled artisans as jade had a status-value exceeding that of gold or silver. Jade became a favourite material for the crafting of Chinese scholars' objects, such as rests for calligraphy brushes, as well as the mouthpieces of some opium pipes, due to the belief that breathing through jade would bestow longevity upon smokers who used such a pipe.

Jadeite, with its bright emerald-green, pink, lavender, orange and brown colours was imported from Burma to China only after about 1800. The vivid green variety became known as Feicui or Kingfisher (feathers) Jade. It quickly became almost as popular as nephrite and a favourite of Qing Dynasty's nouveau riche, while scholars still had strong attachment to nephrite (white jade, or Khotan), which they deemed to be the symbol of a nobleman.

In the history of the art of the Chinese empire, jade has had a special significance, comparable with that of gold and diamonds in the West. Jade was used for the finest objects and cult figures, and for grave furnishings for high-ranking members of the imperial family. Due to that significance and the rising middle class in China, today the finest jade when found in nuggets of "mutton fat" jade — so-named for its marbled white consistency — can fetch $3,000 an ounce, a tenfold increase from a decade ago.

### Prehistoric and Historic India

The Jainist temple of Kolanpak in the Nalgonda district, Andhra Pradesh, India is home to a 5-foot (1.5 m) high sculpture of Mahavira that is carved entirely out of jade. It is the largest sculpture made from a single jade rock in the world. India is also noted for its craftsman tradition of using large amounts of green serpentine or *false jade* obtained primarily from Afghanistan in order to fashion jewellery and ornamental items such as sword hilts and dagger handles.

### Prehistoric and Early Historic Korea

The use of jade and other greenstone was a long-term tradition in Korea (c. 850 BC – AD 668). Jade is found in small numbers of pit-houses and burials. The craft production of small comma-shaped and tubular "jades" using materials such as jade, microcline, jasper, etc., in southern Korea originates from the Middle Mumun Pottery Period (c. 850–550 BC). Comma-shaped jades are found on some of the gold crowns of Silla royalty (c. 300/400–668 AD) and sumptuous elite burials of the Korean Three Kingdoms. After the state of Silla united the Korean Peninsula in 668, the widespread popularisation of death rituals related to Buddhism resulted in the decline of the use of jade in burials as prestige mortuary goods.

### Mâori

Nephrite jade in New Zealand is known as *pounamu* in the Mâori language (often called "greenstone" in New Zealand English), and plays an important role in Mâori culture. It is considered a *taonga*, or treasure, and therefore protected under the Treaty of Waitangi, and the exploitation of it is restricted and closely monitored. It is found only in the South Island of New Zealand, known as *Te Wai Pounamu* in Mâori—"The [land of] Greenstone Water", or *Te Wahi Pounamu*—"The Place of Greenstone".

Tools, weapons and ornaments were made of it; in particular adzes, the 'mere' (short club), and the Hei-tiki (neck pendant). These were believed to have their own mana, handed down as valuable heirlooms, and often given as gifts to seal important agreements. Nephrite jewellery of Maori design is widely popular with locals and tourists, although some of the jade used for these is now imported from British Columbia and elsewhere.

### Canada

***Mesoamerica:*** Jade was a rare and valued material in pre-Columbian Mesoamerica. The only source from which the various indigenous cultures, such as the Olmec and Maya, could obtain jade was located in the Motagua River valley in Guatemala. Jade was largely an elite good, and was usually carved in various ways, whether serving as a medium upon which hieroglyphs were inscribed, or shaped into symbolic figurines. Generally, the material was highly symbolic, and it was often employed in the performance of ideological practices and rituals.

### *Enhancement*

Jade may be enhanced (sometimes called "stabilized"). Note that some merchants will refer to these as Grades, but it is important to bear in mind that degree of enhancement is different from colour and texture quality. In other words, Type A jadeite is not enhanced but can have poor colour and texture. There are three main methods of enhancement, sometimes referred to as the ABC Treatment System:

- Type A jadeite has not been treated in any way except surface waxing.
- Type B treatment involves exposing a promising but stained piece of jadeite to chemical bleaches and/or acids and impregnating it with a clear polymer resin. This results in a significant improvement of transparency and colour of the material. Currently, infrared spectroscopy is the most accurate test for the detection of polymer in jadeite.
- Type C jade has been artificially stained or dyed. The effects are somewhat uncontrollable and may result in a dull brown. In any case, translucency is usually lost.
- B+C jade is a combination of B and C: it has been both impregnated and artificially stained.
- Type D jade refers to a composite stone such as a doublet comprising a jade top with a plastic backing.

## Kanoite

Kanoite is a light pinkish brown silicate mineral that is found in metamorphic rocks. It is an inosilicate and has a formula of $(Mg,Mn^{2+})_2Si_2O_6$. It is a member of pyroxene group and clinopyroxene subgroup.

### *Crystallography*

Kanoite crystallizes in the monoclinic crystal system. Its Hermann–Mauguin Symbol is 2/m. Under this crystal system, the three axes of the crystal are all different in length. The a and the b axes are perpendicular, and b and c axes are perpendicular. The a and c axes make an oblique shape. The axial ratio for kanoite is a:b:c =1.0894:1:0.5884 and the cell dimensions are: a = 9.73, b = 8.93 and c = 5.26 Å with Z = 4. Kanoite has a 2-fold axis and a mirror plane.

Kanoite is birefringent. It occurs as a mineral has 3 different indices of refraction. When the light passes through the Kanoite medium, the light splits due to unequal reflection from the crystal faces. As kanoite is birefringent, it is also anisotropic. In an anisotropic mineral, the velocity of light differs as the direction of the crystal changes.

### *Discovery and Occurrence*

Kanoite is a rare mineral which was found in Tatehira mine, Kumaishi, Oshima Peninsula, Hokkaido, Japan in 1977. In the type locality kanoite occurs along a joint that cuts a pyroxmangite-cummingtonite metamorphic rock in a manganese ore deposit. The region has undergone contact metamorphism as magma intruded the area. It was named to honour Hiroshi Kano, a petrology professor at Akita University in Japan.

It has also been reported from Broken Hill, New South Wales, Australia, the Semail Ophiolite in Oman, and the Balmat-Edwards Zinc District, Saint Lawrence County, New York.

## Laplandite-(Ce)

Laplandite has a general formula of $Na_4CeTiPO_4Si_7O_{18}\cdot 5H_2O$, and is found primarily in igneous rocks. This silicate mineral has been found as inclusions in pegmatites, primarily in the Kola Peninsula in Lappland, where the mineral's name gets its origin. Laplandite is orthorhombic, which states that crystallographically, it contains three axes of unequal lengths that all intersect at 90 degrees, perpendicular to one another. The shape of the crystal is bipyramidal, and is similar in structure to olivine or aragonite. Because of these different axes lengths, it shows anisotropism, which will allow for the visibility of birefringence. This property can give the mineral very distinct colours when viewed under cross-polarization. Laplandite has three different indices of refraction, which are measures of the speed of light in a vacuum divided by the speed of light within the mineral, determined individually on each axis. Due to these different indices, Laplandite is a biaxial mineral, which states that the mineral will have two optic axes. Under the microscope, this mineral has moderate relief, which describes the contrast between Laplandite's refractive index and the refractive index of the mounting medium on which it is placed. The relief can be seen physically as how easily you can see the boundary lines of the mineral under plane polarized light in a petrographic microscope.

Because of Laplandite's sodium solubility, it has been designated as a candidate for extracting soda from the rocks in which it is found. This readily soluble mineral has given geologists clues about the history of the source of the parent rock, as water-soluble minerals do not form near surface temperatures and pressures. Also, the formation of this type of chemical assemblage is commonly located near phosphate and rare-element deposits, giving it another important characteristic as a good indicator to where more economic minerals can be found.

### Larimar

Larimar, also called "Stefilia's Stone", is a rare blue variety of pectolite found only in the Dominican Republic, in the Caribbean. Its colouration varies from white, light-blue, green-blue to deep blue.

### History

The Dominican Republic's Ministry of Mining records show that on 23 November 1916 Father Miguel Domingo Fuertes Loren of the Barahona Parish requested permission to explore and exploit the mine of a certain blue rock he had discovered. Pectolites were not yet known in the Dominican Republic and the request was rejected.

In 1974, at the foot of the Bahoruco Range, the coastal province of Barahona, Miguel Méndez and Peace Corps volunteer Norman Rilling *rediscover* Larimar on a beach. Natives, who believed the stone came from the sea, called the gem *Blue Stone*. Miguel took his young daughter's name Larissa and the Spanish word for sea (*mar*) and formed *Larimar*, by the colours of the water of the Caribbean Sea, where it was found. The few stones they found were alluvial sediment, washed into the sea by the Bahoruco River. An upstream search revealed the in situ outcrops in the range and soon the *Los Chupaderos* mine was formed.

### Geology

Larimar is a type of pectolite, or a rock composed largely of pectolite, an acid silicate hydrate of calcium and sodium. Although pectolite is found in many locations, none have the unique *volcanic blue* colouration of larimar. This blue colour, distinct from that of other pectolites, is the result of copper substitution for calcium.

Miocene volcanic rocks, andesites and basalts, erupted within the limestones of the south coast of the island. These rocks contained cavities or vugs which were later filled with a variety of minerals including the blue pectolite. These pectolite cavity fillings are a secondary occurrence within the volcanic flows, dikes and plugs. When these rocks erode the pectolite fillings are carried downslope to end up in the alluvium and the beach gravels. The Bahoruco River carried the pectolite bearing sediments to the sea. The tumbling action along the streambed provided the natural polishing to the blue larimar which makes them stand out in contrast to the dark gravels of the streambed.

### *Los Chupaderos*

The most important outcrop of blue pectolite is located at *Los Chupaderos*, in the section of *Los Checheses*, about 10 kilometres southwest of the city of Barahona, in the south-western region of the Dominican Republic. It is a single mountainside now perforated with approximately 2,000 vertical shafts, surrounded by rainforest vegetation and deposits of blue-coloured mine tailings.

### *Jewelry*

Larimar jewelry is offered to the public in the Dominican Republic, and elsewhere in the Caribbean as a local speciality. Most jewelry produced is set in silver, but sometimes high-grade larimar is also set in gold. It also has become available elsewhere. Some Far-East manufacturer have started to use it in their production and buy large quantities of raw stones as long as this is still permitted.

Quality grading is according to colouration and the typical mineral crystal configuration in the stone. Larimar also comes in green and even with red spots, brown strikes etc. due to other matters and / or oxidation. But the more intense the blue, and the contrasts in the stone, the higher and rarer is the quality. The blue colour is photosensitive and fades with time if exposed to too much light and heat.

## Nambulite

The general formula for nambulite is $(Li,Na)Mn_4Si_5O_{14}(OH)$. It is named after the mineralogist, Professor Matsuo Nambu (born 1917) of Tohoko University, Japan, who is known for his research in manganese minerals. The mineral was first discovered in the Funakozawa Mine, northeastern Japan, a metasedimentary manganese ore.

Nambulite is formed from the reaction between a hydrothermal solution and rhodonite, and commonly creates veins in the host rock. Other than a collector's gem, however, it has little economic value.

It belongs to the crystal system Triclinic-Pinacoidal (or Triclinic-Normal), meaning that it has three axes of unequal length (a, b, c), all intersecting at oblique angles with each other (none of the angles are equal to 90°). It belongs to the crystal class 1, meaning that any point on the crystal that is rotated 360° and then completely inverted will meet with an equal (but opposite) point on the crystal. It's space group is P 1.

The three axes (a, b, c) have different indices of refraction, $n_a$=1.707, $n_b$=1.710, $n_c$=1.730. The index of refraction (RI) can be defined as n = cair/cmineral, where "n" is the index of refraction and "c" is the speed of light. The maximum birefringence is .023, the difference between the highest ($n_c$=1.730) and lowest ($n_a$=1.707) indices of refraction within the mineral.

In a medium with an index of refraction equaling 1.53, Nambulite has a calculated relief of 1.71-1.73, giving it a moderate to high relief. Relief is a measure of the difference between the index of refraction of the mineral and that of the medium (often Canada balsam or other epoxy with an RI of around 1.53-1.54).

Nambulite is an *anisotropic* crystal, where the velocity of light that passes through the crystal varies depending on the crystallographic direction. In contrast, an *isotropic* crystal includes all isometric crystals, and the velocity of light is equal in all directions. The mineral exhibits slight pleochroism. Pleochroism is an optical property observed when the mineral is viewed under the microscope in plane polarized light, and when it the stage of the microscope is rotated the observed colours change. The colour change is due to different wavelengths being absorbed in different directions, and the colour of the mineral depends on the crystallographic orientation.

## *Omphacite*

Omphacite is a member of the pyroxene group of silicate minerals with formula: (Ca, Na)(Mg, $Fe^{2+}$, Al)$Si_2O_6$. It is a variably deep to pale green or nearly colourless variety of pyroxene. Omphacite compositions are intermediate between calcium-rich augite and sodium-rich jadeite. It crystallizes in the monoclinic system with prismatic, typically twinned forms, though usually anhedral. Its space group (P2/n) is distinct from that of augite and jadeite (C2/c). It exhibits the typical near 90° pyroxene cleavage. It is brittle with specific gravity of 3.29 to 3.39 and a Mohs hardness of 5 to 6.

It is a major mineral component of eclogite along with pyrope garnet and also occurs in blueschist facies and UHP (ultrahigh-pressure) metamorphic rocks. It also occurs in eclogite xenoliths from kimberlite as well as in crustal rocks metamorphosed at high pressures. Associated minerals in eclogites include garnet, quartz or coesite, rutile, kyanite, phengite, and lawsonite. Minerals such as glaucophane, lawsonite, titanite, and epidote occur with omphacite in blueschist facies metamorphic rocks. The name "jade," usually referring to rocks made of jadeite, is sometimes also applied to rocks consisting entirely of omphacite.

The name "omphacite" has been applied to compositions that contain between 20% and 80% jadeite. The stability of intermediate compositions between augite and jadeite is not well-understood, but miscibility gaps appear to be present at temperatures below 300 °C to 400 °C, and perhaps at higher temperatures. Pairs of pyroxenes — both augite plus omphacite and omphacite plus jadeite — appear to have existed in equilibrium at low temperatures.

It was first described in 1815 in the Münchberg Metamorphic complex, Franconia, Bavaria, Germany. The name *omphacite* derives from the Greek *omphax* or *unripe grape* for the typical green colour.

### *Pyroxene*

The pyroxenes are a group of important rock-forming inosilicate minerals found in many igneous and metamorphic rocks. They share a common structure consisting of single chains of silica tetrahedra and they crystallize in the monoclinic and orthorhombic systems. Pyroxenes have the general formula $XY(Si,Al)_2O_6$ (where X represents calcium, sodium, iron$^{+2}$ and magnesium and more rarely zinc, manganese and lithium and Y represents ions of smaller size, such as chromium, aluminium, iron$^{+3}$, magnesium, manganese, scandium, titanium, vanadium and even iron$^{+2}$). Although aluminium substitutes extensively for silicon in silicates such as feldspars and amphiboles, the substitution occurs only to a limited extent in most pyroxenes.

The name pyroxene comes from the Greek words for *fire* and *stranger*. Pyroxenes were named this way because of their presence in volcanic lavas, where they are sometimes seen as crystals embedded in volcanic glass; it was assumed they were impurities in the glass, hence the name "fire strangers". However, they are simply early-forming minerals that crystallized before the lava erupted.

The upper mantle of Earth is composed mainly of olivine and pyroxene. A piece of the mantle is shown at right (orthopyroxene is black, diopside (containing chromium) is bright green, and olivine is yellow-green) and is dominated by olivine, typical for common peridotite. Pyroxene and feldspar are the major minerals in basalt and gabbro.

## Chemistry and Nomenclature of the Pyroxenes

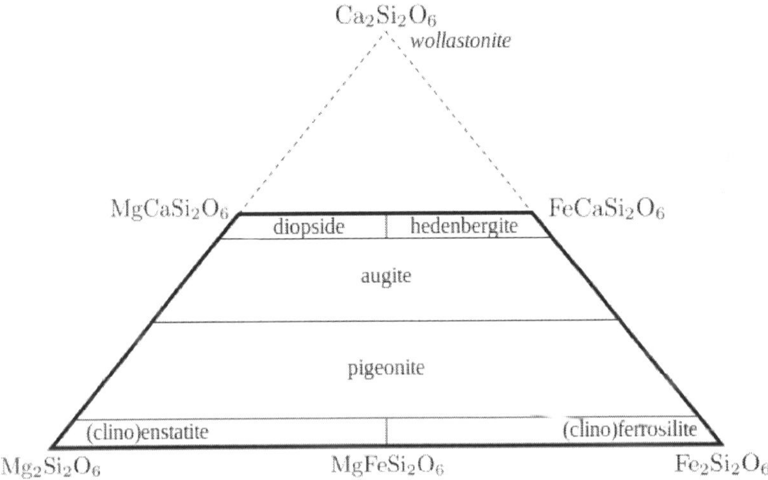


***Figure:*** *The nomenclature of the calcium, magnesium, iron pyroxenes.*

The chain silicate structure of the pyroxenes offers much flexibility in the incorporation of various cations and the names of the pyroxene minerals are primarily defined by their chemical composition. Pyroxene minerals are named according to the chemical species occupying the X (or M2) site, the Y (or M1) site, and the tetrahederal T site. Cations in Y (M1) site are closely bound to 6 oxygens in octahedral coordination. Cations in the X (M2) site can be coordinated with 6 to 8 oxygen atoms, depending on the cation size. Twenty mineral names are recognised by the International Mineralogical Association's Commission on New Minerals and Mineral Names and 105 previously used names have been discarded (Morimoto *et al.*, 1989).

A typical pyroxene has mostly silicon in the tetrahedral site and predominately ions with a charge of +2 in both the X and Y sites, giving the approximate formula $XYT_2O_6$. The names of the common calcium – iron – magnesium pyroxenes are defined in the 'pyroxene quadrilateral' shown in figure. The enstatite-ferrosilite series

($[Mg,Fe]SiO_3$) contain up to 5 mol.% calcium and exists in three polymorphs, orthorhombic orthoenstatite and protoenstatite and monoclinic clinoenstatite (and the ferrosilite equivalents). Increasing the calcium content prevents the formation of the orthorhombic phases and pigeonite ($[Mg,Fe,Ca][Mg,Fe]Si_2O_6$) only crystallises in the monoclinic system. There is not complete solid solution in calcium content and Mg-Fe-Ca pyroxenes with calcium contents between about 15 and 25 mol.% are not stable with respect to a pair of exolved crystals. This leads to a miscibility gap between pigeonite and augite compositions. There is an arbitrary separation between augite and the diopside-hedenbergite ($CaMgSi_2O_6$ – $CaFeSi_2O_6$) solid solution. The divide is taken at >45 mol.% Ca. As the calcium ion cannot occupy the Y site, pyroxenes with more than 50 mol.% calcium are not possible. A related mineral wollastonite has the formula of the hypothetical calcium end member but important structural differences mean that it is not grouped with the pyroxenes.

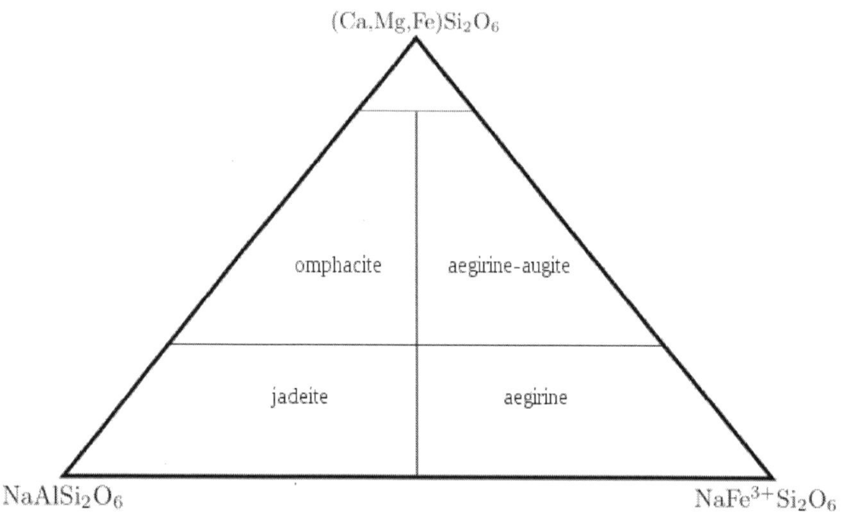


***Figure:*** *The nomenclature of the sodium pyroxenes.*

Magnesium, calcium and iron are by no means the only cations that can occupy the X and Y sites in the pyroxene structure. A second important series of pyroxene minerals are the sodium-rich pyroxenes, corresponding to nomenclature shown in figure. The inclusion of sodium, which has a charge of +1, into the pyroxene implies the need for a mechanism to make up the "missing" positive charge. In jadeite and aegirine this is added by the inclusion of a +3 cation (aluminium and iron(III) respectively) on the Y site. Sodium pyroxenes with more than 20 mol.% calcium, magnesium or iron(II) components are known as omphacite and aegirine-augite, with 80% or more of these components the pyroxene falls in the quadrilateral shown in figure.

| ***Table:*** *Order of cation occupation in the pyroxenes* | | | | | | | | | | | | | | | | | |
|---|---|---|---|---|---|---|---|---|---|---|---|---|---|---|---|---|---|
| T | | Si | Al | $Fe^{3+}$ | | | | | | | | | | | | | |
| Y | | | Al | $Fe^{3+}$ | $Ti^{4+}$ | Cr | V | $Ti^{3+}$ | Zr | Sc | Zn | Mg | $Fe^{2+}$ | Mn | | | |
| X | | | | | | | | | | | | Mg | $Fe^{2+}$ | Mn | Li | Ca | Na |

In assigning ions to sites the basic rule is to work from left to right in this table first assigning all silicon to the T site then filling the site with remaining aluminium and finally iron(III), extra aluminium or iron can be accommodated in the Y site and bulkier ions on the X site. Not all the resulting mechanisms to achieve charge neutrality follow the sodium example above and there are several alternative schemes:

1. Coupled substitutions of 1+ and 3+ ions on the X and Y sites respectively. For example Na and Al give the jadeite ($NaAlSi_2O_6$) composition.
2. Coupled substitution of a 1+ ion on the X site and a mixture of equal numbers of 2+ and 4+ ions on the Y site. This leads to *e.g.* $NaFe^{2+}_{0.5}Ti^{4+}_{0.5}Si_2O_6$.
3. The Tschermak substitution where a 3+ ion occupies the Y site and a T site leading to *e.g.* $CaAlAlSiO_6$.

In nature, more than one substitution may be found in the same mineral.

## *Pyroxene Minerals*

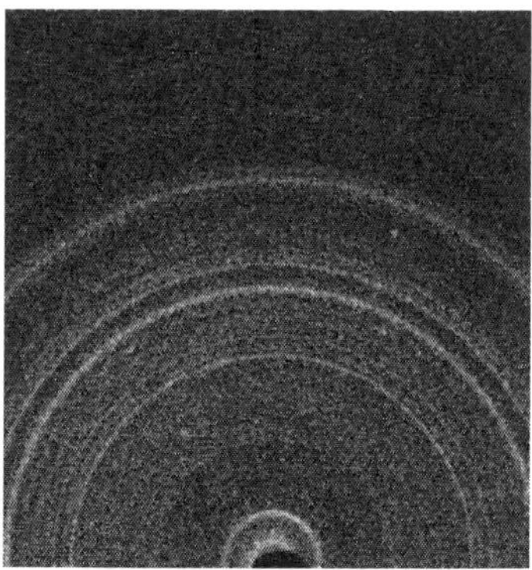

***Figure:*** *First X-ray diffraction view of Martian soil - CheMin analysis reveals feldspar, pyroxenes, olivine and more (Curiosity rover at "Rocknest", October 17, 2012).*

- Clinopyroxenes (monoclinic)
  - o Aegirine (Sodium Iron Silicate)
  - o Augite (Calcium Sodium Magnesium Iron Aluminium Silicate)
  - o Clinoenstatite (Magnesium Silicate)
  - o Diopside (Calcium Magnesium Silicate, $CaMgSi_2O_6$)
  - o Esseneite (Calcium Iron Aluminium Silicate)
  - o Hedenbergite (Calcium Iron Silicate)
  - o Jadeite (Sodium Aluminium Silicate)
  - o Jervisite (Sodium Calcium Iron Scandium Magnesium Silicate)
  - o Johannsenite (Calcium Manganese Silicate)
  - o Kanoite (Manganese Magnesium Silicate)
  - o Kosmochlor (Sodium Chromium Silicate)
  - o Namansilite (Sodium Manganese Silicate)
  - o Natalyite (Sodium Vanadium Chromium Silicate)
  - o Omphacite (Calcium Sodium Magnesium Iron Aluminium Silicate)
  - o Petedunnite (Calcium Zinc Manganese Iron Magnesium Silicate)
  - o Pigeonite (Calcium Magnesium Iron Silicate)
  - o Spodumene (Lithium Aluminium Silicate)
- Orthopyroxenes (orthorhombic)
  - o Hypersthene (Magnesium Iron Silicate)
  - o Donpeacorite, $(MgMn)MgSi_2O_6$
  - o Enstatite, $Mg_2Si_2O_6$
  - o Ferrosilite, $Fe_2Si_2O_6$
  - o Nchwaningite (Hydrated Manganese Silicate)

## Pyroxferroite

Pyroxferroite $(Fe^{2+},Ca)SiO_3$ is a silicate mineral of the pyroxene group. It is mostly composed of iron, silicon and oxygen, with smaller fractions of calcium and several other metals. Together with armalcolite and tranquillityite, it is one of the three minerals which were discovered on the Moon. It was then found in Lunar and Martian meteorites as well as a mineral in the Earth's crust. Pyroxferroite can also be produced by annealing synthetic clinopyroxene at high pressures and temperatures. The mineral is metastable and gradually decomposes at ambient conditions, but this process can last billions of years.

### Etymology

Pyroxferroite is named from *pyroxene* and *ferrum* (Latin for iron), as the iron-rich analogue of pyroxmangite. The word pyroxene, in turn comes from the Greek words for *fire* and *stranger*. Pyroxenes were named this way because of their presence in volcanic lavas, where they are sometimes seen as crystals embedded in volcanic glass; it was assumed they were impurities in the glass, hence the name "fire strangers". However, they are simply early-forming minerals that crystallized before the lava erupted.

### Occurrence

Pyroxferroite was first discovered in 1969 in lunar rock samples from the Sea of Tranquility during the Apollo missions. Together with armalcolite and tranquillityite, it is one of the three minerals which were first found on the Moon. Later, pyroxferroite was detected in Lunar and Martian meteorites recovered in Oman. It also occurs in the Earth's crust, in association with clinopyroxene, plagioclase, ilmenite, cristobalite, tridymite, fayalite, fluorapatite and potassic feldspar, and forms series with pyroxmangite. Pyroxferroite has been found in the Isanago mine, in Kyoto Prefecture, Japan; near Iva, Anderson County, South Carolina, USA; from Vaaster Silfberg, Vaarmland, Sweden; and Lapua, Finland. In the original lunar samples, pyroxferroite was associated with similar minerals, but also with troilite which is rare on Earth, but is common on the Moon and Mars.

### Synthesis

Synthetic pyroxferroite crystals can be produced by compressing synthetic clinopyroxene (composition $Ca_{0.15}Fe_{0.85}SiO_3$) to a pressure in the range of 10–17.5 kbar and heating it to 1130–1250 °C. It is metastable at low temperatures and pressures: at pressures below 10 kbar pyroxferroite converts to a mixture of olivine, pyroxene and a silicon dioxide phase, whereas at low temperatures, it transforms to a clinopyroxene. The presence of cristobalite, vesicular texture and some other petrographic observations indicate that the lunar pyroxferroite was produced upon rapid cooling from low-pressure and high-temperature (volcanic) conditions, i.e. that the mineral is metastable. However, the conversion rate is very slow and pyroxferroite can exist at low temperatures for periods longer than 3 billion years.

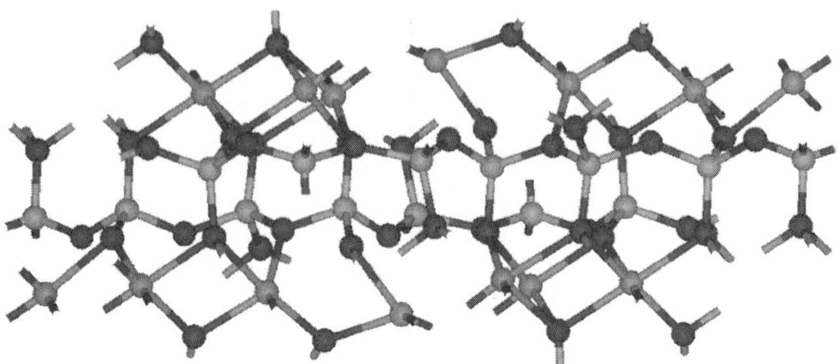

***Figure:*** *Crystal structure. Colours: blue – Fe, gray – Si, red – oxygen.*

### *Properties*

The crystal structure of pyroxferroite contains silicon-oxygen chains with a repeat period of seven $SiO_4$ tetrahedra. These chains are separated by polyhedra where a central metal atom is surrounded by 6 or 7 oxygen atoms; there are 7 inequivalent metal polyhedra in the unit cell. The resulted layers are parallel to (110) planes in pyroxferroite, whereas they are parallel to (100) planes in pyroxenes.

Chemical composition of pyroxferroite can be decomposed into elementary oxides as follows: FeO (concentration 44–48%), $SiO_2$(45–47%), CaO (4.7–6.1%), MnO (0.6–1.3%), MgO (0.3-1%), $TiO_2$ (0.2–0.5%) and $Al_2O_3$ (0.2–1.2%). Whereas magnesium is usually present at about 0.8%, in some samples it had an undetectably low concentration.

## Sapphirine

Sapphirine is a rare mineral, a silicate of magnesium and aluminium with the chemical formula $(Mg,Al)_8(Al,Si)_6O_{20}$ (with iron as a major impurity). Named for its sapphire-like colour, sapphirine is primarily of interest to researchers and collectors: well-formed crystals are treasured and occasionally cut into gemstones. Sapphirine has also been synthesized for experimental purposes via a hydrothermal process.

### *Properties*

Typical colours range from light to dark sapphire blue, bluish to brownish green, green, and bluish or greenish gray to black; less common colours include yellow, pale red, and pink to purplish pink. Sapphirine is relatively hard (7.5 on Mohs scale), usually transparent to translucent, with a vitreous lustre. Crystallising in the monoclinic system, sapphirine is typically anhedral or granular in habit, but may also be tabular or in aggregates: Twinning is uncommon. Fracture is subconchodial to uneven, and there is one direction of perfect cleavage. The specific gravity of sapphirine is 3.54–3.51, and its streak is white.

Sapphirine's refractive index (as measured by monochromatic sodium light, 589.3 nm) ranges from 1.701 to 1.718 with a birefringence of 0.006–0.007, biaxial negative. Refractive index values may correspond to colour: brownish green specimens will possess the highest values, purplish-pink specimens the lowest, and blue specimens will be intermediate between them. Pleochroism may be extreme, with trichroic colours ranging from: colourless, pale yellow or red; sky to lavender blue, or bluish-green; to dark blue. There is no reaction under ultraviolet light.

### Formation and Occurrence

While there is evidence of magmatic origin in some deposits, sapphirine is primarily a product of high grade metamorphism in environments poor in silica and rich in magnesium and aluminium. However, sapphirine occurs in a variety of rocks, including granulite and amphibolite facies, calc-silicate skarns, and quartzites; it is also known from xenoliths. Associated minerals include: calcite, chrysoberyl, cordierite, corundum, garnet, kornerupine, kyanite, phlogopite, scapolite, sillimanite, spinel, and surinamite.

Large crystals of fine clarity and colour are known from very few locales: The Central Province (Hakurutale and Munwatte) of Sri Lanka has long been known as a source of facetable greenish blue to dark blue material, and crystals up to 30 mm or more in size have been found in Fianarantsoa (Betroka District) and Toliara Province (Androy and Anosy regions), southern Madagascar. Sapphirine's type locality is Fiskenaesset (Fiskenaes), Nuuk region, western Greenland, which is where the mineral was discovered in 1819.

Other notable localities include: Western Hoggar, Algeria; the Napier complex of Enderby Land and the Vestfold Hills of Antarctica; Delegate, New South Wales and the Strangways Range of the Northern Territory, Australia; Wilson Lake, Labrador; Donghai, Jiangsu province, China; Kittilä, Lapland, Finland; Ariège, Midi-Pyrénées, France; Waldheim, Saxony, Germany; Dora Maira Massif, Province of Cuneo, Piedmont, Italy; Ulstein, Møre og Romsdal, and Meløy, Nordland, Norway; the Messina District of Limpopo Province and the Okiep Copper District of Northern Cape Province, South Africa; Falkenberg Municipality, Halland County, Sweden; Mautia Hill in the

Kongwa region of Central Province, Tanzania; Isle of Harris, Outer Hebrides, Scotland; the Bani Hamid area of Semail Ophiolite, United Arab Emirates; the Dome Rock Mountains of La Paz County, Arizona, Stockdale, Riley County, Kansas; Cortlandt, New York; and Clay County, North Carolina.

## Spodumene

Spodumene is a pyroxene mineral consisting of lithium aluminium inosilicate, $LiAl(SiO_3)_2$, and is a source of lithium. It occurs as colourless to yellowish, purplish, or lilac kunzite, yellowish-green or emerald-green hiddenite, prismatic crystals, often of great size. Single crystals of 14.3 m (47 ft) in size are reported from the Black Hills of South Dakota, United States.

The normal low-temperature form α-spodumene is in the monoclinic system whereas the high-temperature β-spodumene crystallizes in the tetragonal system. The normal α-spodumene converts to β-spodumene at temperatures above 900 °C. Crystals are typically heavily striated parallel to the principal axis. Crystal faces are often etched and pitted with triangular markings.

### *Discovery and Occurrence*

Spodumene was first described in 1800 for an occurrence in the type locality in Utö, Södermanland, Sweden. The name is derived from the Greek *spodumenos* (σπïäõμåíïò), meaning "burnt to ashes," owing to the opaque, ash-grey appearance of material refined for use in industry.

Spodumene occurs in lithium-rich granite pegmatites and aplites. Associated minerals include: quartz, albite, petalite, eucryptite, lepidolite and beryl. Transparent material has long been used as a gemstone with varieties kunzite and hiddenite noted for their strong pleochroism. Source localities include Afghanistan, Australia, Brazil, Madagascar, Pakistan, Québec in Canada and North Carolina, California in the USA.

### *Economic Importance*

Spodumene is an important source of lithium for use in ceramics, mobile phone and automotive batteries, medicine and as a fluxing agent. Lithium is extracted from spodumene by fusing in acid.

World production of lithium via spodumene is around 80,000 metric tonnes per annum, primarily from the Greenbushes pegmatite of Western Australia, and some Chinese and Chilean sources. The

Talison mine in Greenbushes, Western Australia has an estimated reserve of 13 million tonnes. Some think that spodumene will become a less important source of lithium due to the emergence of alkaline brine lake sources in Chile, China and Argentina, which produce lithium chloride directly. Lithium chloride is converted to lithium carbonate and lithium hydroxide by reaction with sodium carbonate and calcium hydroxide respectively.

But, pegmatite-based projects benefit from being quicker to move into production than brines, which can take 18 months to 3 years, depending on evaporation rates. With pegmatites, once a mill is built, the production of lithium carbonate is only a matter of days. Another key advantage that spodumene has over its more popular brine rivals, is the purity of the lithium carbonate it can produce. While all product used by the battery industry have to grade at least 99.5% lithium carbonate, the make up of that final 0.5% is important. If it contains higher amounts of iron, magnesium or other deleterious materials it is less attractive to end users.

### *Gemstone Varieties*

***Hiddenite:*** Hiddenite is a pale emerald green gem variety first reported from Alexander County, North Carolina, U. S. A.

This emerald green variety of spodumene is coloured by chromium, just like emeralds. Not all green spodumene is coloured with chromium, which tend to have a lighter colour, and therefore are not true hiddenite.

### *Kunzite*

Kunzite is a pink to lilac coloured gemstone, a variety of spodumene with the colour coming from minor to trace amounts of manganese. Some (but not all) kunzite used for gemstones has been heated to enhance its colour. It is also frequently irradiated to enhance the colour. Many kunzites fade when exposed to sunlight.

Kunzite was discovered in 1902, and was named after George Frederick Kunz, Tiffany & Co's chief jeweler at the time, and a noted mineralogist. It has been found in Brazil, USA, Canada, CIS, Mexico, Sweden, Western Australia, Afghanistan and Pakistan.

One notable example of kunzite used in jewellery is in the Russian Palmette tiara and necklace worn by the Duchess of Gloucester.

## Tschermakite

The endmember hornblende tschermakite ($Ca_2(Mg,Fe^{2+})_3Al_2(Si_6Al_2)O_{22}(OH)_2$) is a calcium rich monoclinic amphibole mineral.

It is frequently synthesized along with its ternary solid solution series members tremolite and cummingtonite so that the thermodynamic properties of its assemblage can be applied to solving other solid solution series from a variety of amphibole minerals.

### *Mineral Composition*

Tschermakite is an end-member of the hornblende subgroup in the calcic-amphibole group. Calcium-rich amphiboles have the general formula $X_{2\text{-}3}\ Y_5\ Z_8\ O_{22}\ (OH)_2$ where X=Ca, Na, K, Mn; Y=Mg, $Fe^{+2}$, $Fe^{+3}$, Al, Ti, Mn, Cr, Li, Zn; Z=Si, Al (Deer et al., 1963). The structure of tremolite ($Ca_2Mg_5(Si_8O_{22})(OH,F)_2$), another calcic amphibole, is commonly used as the standard for calcic amphiboles from which the formulae for their substitutions are derived. The wide range in variety of minerals classified in the amphibole group is due to its great ability for ionic replacement resulting in a widely varying chemical composition. Amphiboles can be classified on the basis of the substitution of ions on the X site as well as the substitution of AlAl for Si(Mg, $Fe^{+2}$). In the calcium amphiboles like tschermakite $Ca_2(Mg, Fe^{2+})_3Al_2\ (Si_6\ Al_2)\ O_{22}(OH)_2$, the predominant ion in the X position is occupied by Ca as in tremolite, while the substitution MgSi<->AlAl occurs on the Y and the tetrahedral Z site.

### *Geologic Occurrence*

Hornblendes are the most common of the amphiboles and are formed in a wide range of Pressure-Temperature environments. Tschermakite is found in eclogites and ultramafic igneous rocks as well as in medium to high-grade metamorphic rocks. The mineral is widespread throughout the world but has most notably been studied in Greenland, Scotland, Finland, France, and the Ukraine (Anthony, 1995). Because amphibole minerals like Tschermakite are hydrous (contain an OH group), they can break down to denser anhydrous minerals like pyroxene or garnet at high temperatures. Conversely, amphiboles can be recomposed from pyroxenes as a result of crystallizing igneous rocks as well as during metamorphism (Léger and Ferry, 1991). Because of this important quality, P-T conditions have repeatedly been calculated for the crystallization of hornblendes in calc-alkaline magmas (Féménias et al., 2006). In addition to studying tschermakitic content in its natural occurrences, geologists have frequently synthesized this mineral in order to further calculate its place as an endmember hornblende.

### *Namesake Biography*

In 1872 Professor Tschermak founded one of Europe's oldest geoscience journals Mineralogische Mitteilungen (Deu: Mineralogical

Disclosures) or Mineralogy and Petrology. In the first volume of Min. Mitt., Tschermak established some of the early classifications of the amphibole group in relation to the pyroxene group of minerals (Tschermak 1871), which no doubt led to the formula $Ca_2Mg_3Al_4Si_6O_{22}(OH)_2$ being known as the Tschermak molecule, this mineral formula was later assigned the name tschermakite as first proposed by Winchell (1945). Professor Tschermak spent many years working as curator for the Imperial Mineralogical Cabinet. The Mineralogical Dept. of the Imperial Natural History Museum in Vienna – an impressive mineral, meteorite and fossil collection has Professor Tschermak to thank for his detailed inventory system that has helped preserve it to this day as well as the expansion of their meteorite collection. He was a full professor of mineralogy and petrography at the University of Vienna as well as a full member of the Imperial Academy of Sciences in Vienna. He was also the first president of the Viennese (now Austrian) Mineralogical Society, founded in 1901. An obituary for "Hofrat Professor Dr. Gustav Tschermak" written by Edward S. Dana (1927) can be found in the 12th volume of American Mineralogist where Dana recalls the two young scientists earlier work together in the Vienna Mineral Cabinet and remarks on Professor Tschermak's vigor and clarity of mind maintained up to his final days. Gustav Tschermak's third child, Erich von Tschermak-Seysenegg (1871-1962) was a renowned botanist who is credited for independently resdiscovering Gregor Mendel's genetic laws of inheritance by working with similar plant breeding experiments.

### Mineral Structure

The amphibole group consists of an orthorhombic and monoclinic series – hornblendes and tschermakite both belong to the latter crystal structure. The crystal group of tschermakite is 2/m.

Tschermakite and all the hornblende varieties are inosilicates, and like the other rock forming amphiboles are double chain silicates (Klein and Hurlbut, 1985). The amphibole structure is characterized by its two double chains of $SiO_4$ tetrahedra (T1 and T2) sandwiching in a strip of cations (M1, M2 and M3 octahedra). Much of the discussions and studies of both tschermakite and tremolite have been to resolve the varying cation placements and Al substitutions that seem to occur on all T and M sites (Najorka and Gottschalk, 2003).

### Physical Properties

A hand specimen of tschermakite is green to black in colour; its streak will be greenish white. It can be transparent to translucent and has a vitreous luster. Tschermakite shows the characteristic amphibole perfect cleavage on [110]. Its average density is 3.24, with a hardness of 5-6; its fracture will be brittle to conchoidal. In thin section its optic sign and 2V angle cover a wide range and are not very useful for identification. It shows a distinct pleochroism in browns and greens.

### *Special Characteristics*

Much discussion and experimentation on Tschermakite has been in relation to it being synthesized along with other calcic-amphiboles to determine the stoichiometric and barometric constraints of the various amphibole solid solutions series. Because of the (Mg, Fe, Ca),Si<->Al, Al tschermak cation exchange that is fundamental to not only the amphibole group but also the pyroxenes, micas and chlorites (Najorka and Gottschalk, 2003) (Ishida and Hawthorne, 2006). Tschermakite has been synthesized in numerous experiments along with its ternary solid solution end members tremolite and cummingtonite in order to relate its varying compositions to a specific P and T. The thermodynamic data that results from these tests helps to calculate further geothermobarometric equations in both synthesized and natural forms of a variety of minerals.

## Wollastonite

Wollastonite is a calcium inosilicate mineral ($CaSiO_3$) that may contain small amounts of iron, magnesium, and manganese substituting for calcium. It is usually white. It forms when impure limestone or dolostone is subjected to high temperature and pressure sometimes in the presence of silica-bearing fluids as in skarns or contact metamorphic rocks. Associated minerals include garnets, vesuvianite, diopside, tremolite, epidote, plagioclase feldspar, pyroxene and calcite. It is named after the English chemist and mineralogist William Hyde Wollaston (1766–1828).

Some of the properties that make wollastonite so useful are its high brightness and whiteness, low moisture and oil absorption, and low volatile content. Wollastonite is used primarily in ceramics, friction products (brakes and clutches), metalmaking, paint filler, and plastics.

Despite its chemical similarity to the compositional spectrum of the pyroxene group of minerals—where magnesium and iron substitution for calcium ends with diopside and hedenbergite respectively—it is structurally very different, with a third $SiO_4$ tetrahedron in the linked chain (as opposed to two in the pyroxenes).

### *Production Trends*

World production data for wollastonite is not available for many countries and those that are available frequently are 2 to 3 years old. Estimated world production of crude wollastonite ore was in the range of 530,000 to 550,000 tonnes in 2010. World reserves of wollastonite were estimated to exceed 90 million tonnes, with probable reserves of about 270 million tonnes. However, many large deposits have not been surveyed yet.

In 2010, the major producers were China (300,000 tonnes), India (120,000 t), United States (67,000 t), Mexico (30,000 t) and Finland (16,000). Finland has long been a major European supplier of wollastonite, but in 2003 it was joined by Spain with comparable production volumes. In the United States, wollastonite is mined in Willsboro, New York and Gouverneur, New York. Deposits have also been mined commercially in North Western Mexico.

### *Uses*

Wollastonite has industrial importance worldwide. It is used in many industries, mostly by tile factories which have incorporated it into the manufacturing of ceramic to improve many aspects, and this is due to its fluxing properties, freedom from volatile constituents, whiteness, and acicular particle shape. In ceramics, wollastonite decreases shrinkage and gas evolution during firing, increases green and fired strength, maintains brightness during firing, permits fast firing, and reduces crazing, cracking, and glaze defects. In metallurgical applications, wollastonite serves as a flux for welding, a source for calcium oxide, a slag conditioner, and to protect the surface of molten metal during the continuous casting of steel. As an additive in paint, it improves the durability of the paint film, acts as a pH buffer, improves its resistance to weathering, reduces gloss, reduces pigment consumption, and acts as a flatting and suspending agent. In plastics, wollastonite improves tensile and flexural strength, reduces resin consumption, and improves thermal and dimensional stability at elevated temperatures. Surface treatments are used to improve the adhesion between the wollastonite and the polymers to which it is added. As a substitute for asbestos in floor tiles, friction products, insulating board and panels, paint, plastics, and roofing products, wollastonite is resistant to chemical attack, inert, stable at high temperatures, and improves flexural and tensile strength. In some industries, it is used in different percentages of impurities, such as its use as a fabricator of mineral wool insulation, or as an ornamental building material.

Plastics and rubber applications were estimated to account for 25% to 35% of U.S. sales in 2009, followed by ceramics with 20% to 25%; paint, 10% to 15%; metallurgical applications, 10% to 15%; friction products, 10% to 15%; and miscellaneous, 10% to 15%. Ceramic applications probably account for 30% to 40% of wollastonite sales worldwide, followed by polymers (plastics and rubber) with 30% to 35% of sales, and paint with 10% to 15% of sales. The remaining sales were for construction, friction products, and metallurgical applications. The price of raw wollastonite varied in 2008 between US$80 and US$500 per tonne depending on the country and size and shape of the powder particles.

### Substitutes

The acicular nature of many wollastonite products allows it to compete with other acicular materials, such as ceramic fibre, glass fibre, steel fibre, and several organic fibres, such as aramid, polyethylene, polypropylene, and polytetrafluoroethylene in products where improvements in dimensional stability, flexural modulus, and heat deflection are sought. Wollastonite also competes with several nonfibrous minerals or rocks, such as kaolin, mica, and talc, which are added to plastics to increase flexural strength, and such minerals as barite, calcium carbonate, gypsum, and talc, which impart dimensional stability to plastics. In ceramics, wollastonite competes with carbonates, feldspar, lime, and silica as a source of calcium and silicon. Its use in ceramics depends on the formulation of the ceramic body and the firing method.

### Composition

In a pure $CaSiO_3$, each component forms nearly half of the mineral by weight: 48.3% of CaO and 51.7% of $SiO_2$. In some cases, small amounts of iron (Fe), and manganese (Mn), and lesser amounts of magnesium (Mg) substitute for calcium (Ca) in the mineral formula (*e.g.*, rhodonite). Wollastonite can form a series of solid solutions in the system $CaSiO_3$-$FeSiO_3$, or hydrothermal synthesis of phases in the system $MnSiO_3$-$CaSiO_3$.

### Geologic Occurrence

Wollastonite usually occurs as a common constituent of a thermally metamorphosed impure limestone, it also could occur when the silicon is due to metamorphism in contact altered calcareous sediments, or to contamination in the invading igneous rock. In most of these occurrences it is the result of the following reaction between calcite and silica with the loss of carbon dioxide:

$$CaCO_3 + SiO_2 \rightarrow CaSiO_3 + CO_2$$

Wollastonite may also be produced in a diffusion reaction in skarn, it develops when limestone within a sandstone is metamorphosed by a dike, which results in the formation of wollastonite in the sandstone as a result of outward migration of Ca.

### Structure

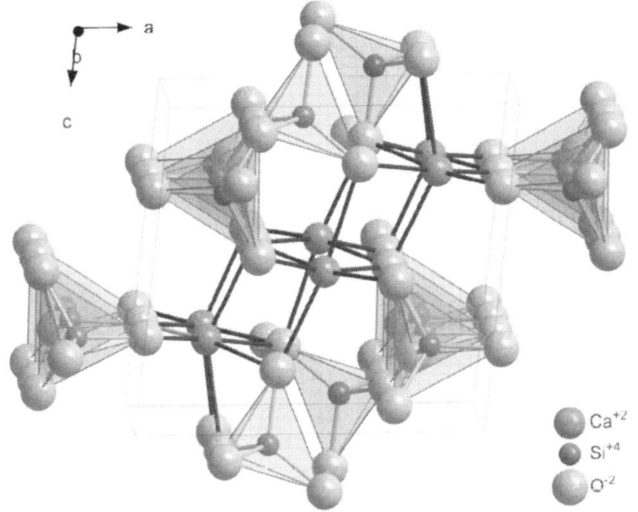


***Figure:*** *Unit cell of triclinic wollastonite-1A*

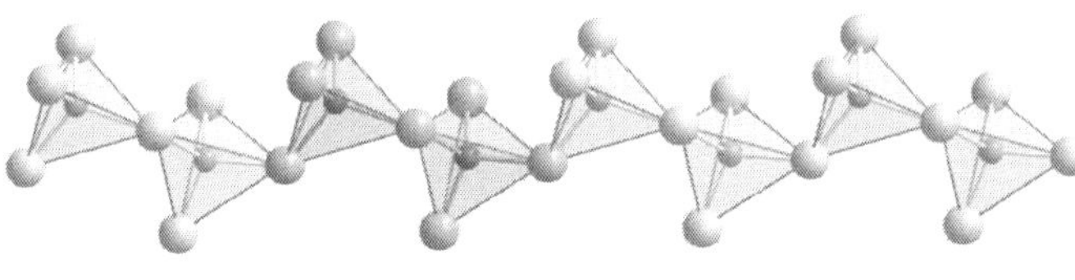


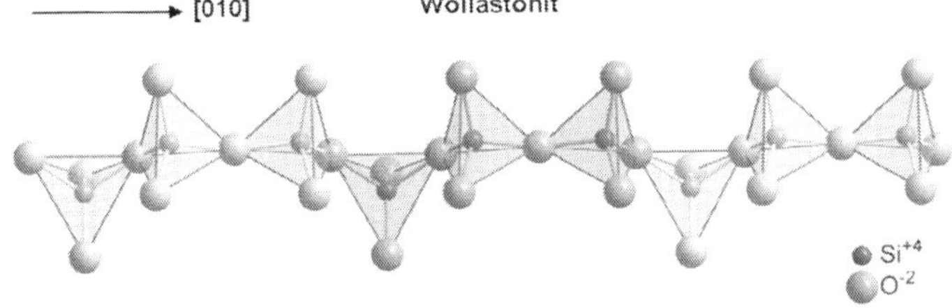


***Figure:*** *Tetrahedra arrangement within the chains in pyroxenes compared to wollastonite*

Wollastonite crystallizes triclinically in space group P1 with the lattice constants $a = 7.94$ Å, $b = 7.32$ Å, c = 7.07 Å; α = 90,03°, β = 95,37°, γ = 103,43° and six formula units per unit cell. Wollastonite was once classed structurally among the pyroxene group, because both of these groups have a ratio of Si:O = 1:3. In 1931, Warren and Biscoe showed that the crystal structure of wollastonite differs from minerals of the pyroxene group, and they classified this mineral within a group known as the pyroxenoids. It has been shown that the pyroxenoid chains are more kinked than those of pyroxene group, and exhibit longer repeat distance. The structure of wollastonite contains infinite chains of [$SiO_4$] tetrahedra sharing common vertices, running parallel to the *b*-axis. The chain motif in wollastonite repeats after three tetrahedra, whereas in pyroxenes only two are needed. The repeat distance in the wollastonite chains is 7.32 Å and equals the length of the crystallographic *b*-axis.

Molten $CaSiO_3$, maintains a tetrahedral $SiO_4$ local structure, at temperatures up to 2000 °C. The nearest neighbour Ca-O coordination decreases from 6.0(2) in the room temperature glass to 5.0(2) in the 1700 °C liquid, coincident with an increasing number of longer Ca-O neighbours.

## Physical and Optical Properties

Wollastonite occurs as bladed crystal masses, single crystals can show an acicular particle shape and usually it exhibits a white colour, but sometimes cream, grey or very pale green.

The streak of wollastonite is white, its Mohs hardness is 4.5–5 and specific gravity is 2.87–3.09. There are more than one cleavage planes for it, there is a perfect cleavage on {100}, good cleavages on {001}, and {102}, and an imperfect cleavage on {101}. It is common for wollastonite to have a twin axis [010], a composition plane (100), and rarely to have a twin axis [001]. The luster is usually vitreous to pearly. The melting point of wollastonite is about 1540 ÚC.

### *Zorite*

Zorite is a silicate mineral with the chemical formula of $Na_2Ti(Si,Al)_3O_9 \cdot nH_2O$. It is named because of its pink colour, after the Russian word "zoria" which refers to the rosy hue of the sky at dawn. It is primarily found in Mount Karnasurta, Lovozero Massif, Kola Peninsula, Russia. The Lovozero Massif is an area with an igneous mountain range, home to various types of minerals such as eudialyte, loparite, and natrosilitite.

Crystallographically, zorite belongs in the orthorhombic group, which has 3 axes, a, b, and c that are of unequal lengths (a“‘b“‘c) that form 90° with each other. It also belongs in the point group 2/m2/m2/m. The state of aggregation for zorite is acicular. Zorite has perfect cleavage along the planes {010} and {001}, while having poor cleavage along the plane {110}. Zorite is anisotropic, which means that the velocity of light is not the same in all directions. It belongs in the biaxial group, because it is an orthorhombic mineral. Under plane polarized light, zorite displays different colours depending on the angle that the light hits the mineral. This quality is called pleochroism and zorite is rose along the x-axis, colourless along the y-axis, and bluish along the z-axis. The index of refraction of zorite is 1.59, which is the velocity of light through vacuum over the velocity of light through zorite. Zorite is studied to better understand silicate structures.

In 2003, zorite was looked into to analyze the symmetry and topology of a family of three minerals found in Russia, Nenadkevichite, Labuntsovite, and Zorite. Zorite was also studied to comprehend how silicate structures change when an element is replaced, for example when the sodium is replaced with potassium, caesium and phosphorus. Furthermore, because of its rarity, zorite is one of the collectors' items coveted for its scarcity, as well as it being a valuable source to understanding silicate topology.

# 9

# Phyllosilicates

## Serpentine Group

The serpentine group are greenish, brownish, or spotted minerals commonly found in serpentinite rocks. They are used as a source of magnesium and asbestos, and as a decorative stone. The name is thought to come from the greenish colour being that of a serpent.

The serpentine group describes a group of common rock-forming hydrous magnesium iron phyllosilicate ($(Mg, Fe)_3Si_2O_5(OH)_4$) minerals; they may contain minor amounts of other elements including chromium, manganese, cobalt or nickel. In mineralogy and gemology, serpentine may refer to any of 20 varieties belonging to the serpentine group. Owing to admixture, these varieties are not always easy to individualize, and distinctions are not usually made. There are three important mineral polymorphs of serpentine: antigorite, chrysotile and lizardite.

The chrysotile group of minerals are polymorphous, meaning that they have the same chemical formulae, but the molecules are arranged into different structures, or crystal lattices. Chrysotile with a fibreous habit is one type of asbestos. Other minerals in the chrysotile group may have a platy habit.

Many types of serpentine have been used for jewellery and hardstone carving, sometimes under the name *false jade* or *Teton jade.*

### *Overview*

Their olive green colour and smooth or scaly appearance is the basis of the name from the Latin *serpentinus*, meaning "serpent rock," according to Best (2003). They have their origins in metamorphic alterations of peridotite and pyroxene. Serpentines may also pseudomorphously replace other magnesium silicates. Alterations may be incomplete, causing physical properties of serpentines to vary widely. Where they form a significant part of the land surface, the soil is unusually high in clay.

Antigorite is the polymorph of serpentine that most commonly forms during metamorphism of wet ultramafic rocks and is stable at the highest temperatures—to over 600 °C at depths of 60 km or so. In contrast, lizardite and chrysotile typically form near the Earth's surface and break down at relatively low temperatures, probably well below 400 °C. It has been suggested that chrysotile is never stable relative to either of the other two serpentine polymorphs.

Samples of the oceanic crust and uppermost mantle from ocean basins document that ultramafic rocks there commonly contain abundant serpentine. Antigorite contains water in its structure, about 13 percent by weight. Hence, antigorite may play an important role in the transport of water into the earth in subduction zones and in the subsequent release of water to create magmas in island arcs, and some of the water may be carried to yet greater depths.

Soils derived from serpentine are toxic to many plants, because of high levels of nickel, chromium, and cobalt; growth of many plants is also inhibited by low levels of potassium and phosphorus and a low ratio of calcium/magnesium. The flora is generally very distinctive, with specialised, slow-growing species. Areas of serpentine-derived soil will show as strips of shrubland and open, scattered small trees (often conifers) within otherwise forested areas; these areas are called serpentine barrens.

Most serpentines are opaque to translucent, light (specific gravity between 2.2–2.9), soft (hardness 2.5–4), infusible and susceptible to acids. All are microcrystalline and massive in habit, never being found as single crystals. Luster may be vitreous, greasy or silky. Colours range from white to grey, yellow to green, and brown to black, and are often splotchy or veined. Many are intergrown with other minerals, such as calcite and dolomite. Occurrence is worldwide; New Caledonia, Canada (Quebec), USA (northern California, Rhode Island, Connecticut, Massachusetts, Maryland and southern Pennsylvania), Afghanistan, Britain (Cornwall and Ireland), Greece (Thessaly), China, Ural Mountains (Russia), France, Korea, Austria (Styria and Carinthia), India (Assam, and Manipur), Myanmar (Burma), New Zealand, Norway and Italy are notable localities. Serpentines find use in industry for a number of purposes, such as railway ballasts, building materials, and the asbestiform types find use as thermal and electrical insulation (chrysotile asbestos). The asbestos content can be released to the air when serpentine is excavated and if it is used as a road surface, forming a long term health hazard by breathing. Asbestos from serpentine can also appear at low levels in water supplies through normal weathering processes, but there is as yet no identified health hazard associated with use or ingestion. In its natural state, some forms of serpentine react with carbon dioxide and re-release oxygen into the atmosphere.

The more attractive and durable varieties (all of antigorite) are termed "noble" or "precious" serpentine and are used extensively as gems and in ornamental carvings. The town of Bhera in the historic Punjab province of the Indian subcontinent was known for centuries for finishing a relatively pure form of green serpentine obtained from quarries in Afghanistan into lapidary work, cups, ornamental sword hilts, and dagger handles. This high-grade serpentine ore was known as *sang-i-yashm* or to the English, *false jade*, and was used for generations by Indian craftsmen for lapidary work. It is easily carved, taking a good polish, and is said to have a pleasingly greasy feel. Less valuable serpentine ores of varying hardness and clarity are also sometimes dyed to imitate jade. Misleading synonyms for this material include "Suzhou jade", "Styrian jade", and "New jade".

New Caledonian serpentine is particularly rich in nickel. The Mâori of New Zealand once carved beautiful objects from local serpentine, which they called *tangiwai*, meaning "tears".

The *lapis atracius* of the Romans, now known as verde antique, or verde antico, is a serpentinite breccia popular as a decorative facing stone. In classical times it was mined at Casambala, Thessaly, Greece. Serpentinite marbles are also widely used: Green *Connemara marble* (or *Irish green marble*) from Connemara, Ireland (and many other sources), and red *Rosso di Levanto marble* from Italy. Use is limited to indoor settings as serpentinites do not weather well.

### Antigorite

Lamellated antigorite occurs in tough, pleated masses. It is usually dark green in colour, but may also be yellowish, gray, brown or black. It has a hardness of 3.5–4 and its lustre is greasy. The monoclinic crystals show micaceous cleavage and fuse with difficulty. Antigorite is named after its type locality, the Valle di Antigorio in Italy.

Bowenite is an especially hard serpentine (5.5) of a light to dark apple green colour, often mottled with cloudy white patches and darker veining. It is the serpentine most frequently encountered in carving and jewellery. The name *retinalite* is sometimes applied to yellow bowenite. The New Zealand material is called *tangiwai*.

Although not an official species, bowenite is the state mineral of Rhode Island: this is also the variety's type locality. A bowenite cabochon featured as part of the "Our Mineral Heritage Brooch", was presented to First Lady Mrs. Lady Bird Johnson in 1967.

Williamsite is a local varietal name for antigorite that is oil-green with black crystals of chromite or magnetite often included. Somewhat resembling fine jade, williamsite is cut into cabochons and beads. It is found mainly in Maryland and Pennsylvania, USA.

Gymnite is an amorphous form of antigorite. It was originally found in the Bare Hills, Maryland, and is named from the Greek, gymnos meaning bare or naked.

### *State Emblem*

In 1965 the California Legislature designated serpentine (the mineral) as "the official State Rock and lithologic emblem."

## Clay Minerals

Clay minerals are hydrous aluminium phyllosilicates, sometimes with variable amounts of iron, magnesium, alkali metals, alkaline earths, and other cations. Clays form flat hexagonal sheets similar to the micas. Clay minerals are common weathering products (including weathering of feldspar) and low temperature hydrothermal alteration products. Clay minerals are very common in fine grained sedimentary rocks such as shale, mudstone, and siltstone and in fine grained metamorphic slate and phyllite.

Clay minerals are usually (but not necessarily) ultrafine-grained (normally considered to be less than 2 micrometres in size on standard particle size classifications) and so may require special analytical techniques for their identification/study. These include x-ray diffraction, electron diffraction methods, various spectroscopic methods such as Mössbauer spectroscopy, infrared spectroscopy, and SEM-EDS or automated mineralogy solutions. These methods can be augmented by polarized light microscopy, a traditional technique establishing fundamental occurrences or petrologic relationships.

Clay minerals can be classified as 1:1 or 2:1, this originates from the fact that they are fundamentally built of tetrahedral silicate sheets and octahedral hydroxide sheets, as described in the structure section below. A 1:1 clay would consist of one tetrahedral sheet and one octahedral sheet, and examples would be kaolinite and serpentine. A 2:1 clay consists of an octahedral sheet sandwiched between two tetrahedral sheets, and examples are talc, vermiculite and montmorillonite.

Clay minerals include the following groups:

- Kaolin group which includes the minerals kaolinite, dickite, halloysite, and nacrite (polymorphs of $Al_2Si_2O_5(OH)_4$).
    - o Some sources include the kaolinite-serpentine group due to structural similarities (Bailey 1980).
- Smectite group which includes dioctahedral smectites such as montmorillonite and nontronite and trioctahedral smectites for example saponite. In 2013, analytical tests by the Curiosity rover found results consistent with the presence of *smectite clay minerals* on the planet Mars.
- Illite group which includes the clay-micas. Illite is the only common mineral.
- Chlorite group includes a wide variety of similar minerals with considerable chemical variation.
- Other 2:1 clay types exist such as sepiolite or attapulgite, clays with long water channels internal to their structure.

Mixed layer clay variations exist for most of the above groups. Ordering is described as random or regular ordering, and is further described by the term reichweite, which is German for range or reach. Literature articles will refer to a R1 ordered illite-smectite, for example. This type would be ordered in an ISISIS fashion. R0 on the other hand describes random ordering, and other advanced ordering types are also found (R3, etc.). Mixed layer clay minerals which are perfect R1 types often get their own names. R1 ordered chlorite-smectite is known as corrensite, R1 illite-smectite is rectorite.

### History

Knowledge of the nature of clay became better understood in the 1930s with advancements in x-ray diffraction technology necessary to analyze the molecular nature of clay particles. Standardization in terminology arose during this period as well with special attention given to similar words that resulted in confusion such as sheet and plane.

### *Structure*

Like all phyllosilicates, clay minerals are characterised by two-dimensional *sheets* of corner sharing $SiO_4$ tetrahedra and/or $AlO_4$ octahedra. The sheet units have the chemical composition $(Al,Si)_3O_4$. Each silica tetrahedron shares 3 of its vertex oxygen atoms with other tetrahedra forming a hexagonal array in two-dimensions. The fourth vertex is not shared with another tetrahedron and all of the tetrahedra "point" in the same direction; i.e. all of the unshared vertices are on the same side of the sheet.

In clays, the tetrahedral sheets are always bonded to octahedral sheets formed from small cations, such as aluminium or magnesium, and coordinated by six oxygen atoms. The unshared vertex from the tetrahedral sheet also forms part of one side of the octahedral sheet, but an additional oxygen atom is located above the gap in the tetrahedral sheet at the centre of the six tetrahedral. This oxygen atom is bonded to a hydrogen atom forming an OH group in the clay structure. Clays can be categorized depending on the way that tetrahedral and octahedral sheets are packaged into *layers*. If there is only one tetrahedral and one octahedral group in each layer the clay is known as a 1:1 clay. The alternative, known as a 2:1 clay, has two tetrahedral sheets with thc unshared vertex of each sheet pointing towards each other and forming each side of the octahedral sheet.

Bonding between the tetrahedral and octahedral sheets requires that the tetrahedral sheet becomes corrugated or twisted, causing ditrigonal distortion to the hexagonal array, and the octahedral sheet is flattened. This minimizes the overall bond-valence distortions of the crystallite.

Depending on the composition of the tetrahedral and octahedral sheets, the layer will have no charge, or will have a net negative charge. If the layers are charged this charge is balanced by interlayer cations such as $Na^+$ or $K^+$. In each case the interlayer can also contain water. The crystal structure is formed from a stack of layers interspaced with the interlayers.

## Mica

The mica group of sheet silicate (phyllosilicate) minerals includes several closely related materials having close to perfect basal cleavage. All are monoclinic, with a tendency towards pseudohexagonal crystals, and are similar in chemical composition. The nearly perfect cleavage, which is the most prominent characteristic of mica, is explained by the hexagonal sheet-like arrangement of its atoms. The word "mica" is derived from the Latin word *mica*, meaning "a crumb", and probably influenced by *micare*, "to glitter".

### Mica Classification

Chemically, micas can be given the general formula

$X_2Y_{4-6}Z_8O_{20}(OH,F)_4$

in which $X$ is K, Na, or Ca or less commonly Ba, Rb, or Cs;

$Y$ is Al, Mg, or Fe or less commonly Mn, Cr, Ti, Li, etc.;

$Z$ is chiefly Si or Al, but also may include $Fe^{3+}$ or Ti.

Structurally, micas can be classed as dioctahedral ($Y = 4$) and trioctahedral ($Y = 6$). If the $X$ ion is K or Na, the mica is a "common" mica, whereas if the $X$ ion is Ca, the mica is classed as a "brittle" mica.

### Trioctahedral Micas

Common micas:

- Biotite
- Lepidolite
- Muscovite
- Phlogopite
- Zinnwaldite

Brittle micas:

- Clintonite

### Interlayer Deficient Micas

Very fine-grained micas, which typically show more variation in ion and water content, are informally termed "clay micas". They include:

- Hydro-muscovite with $H_3O^+$ along with K in the $X$ site;
- Illite with a K deficiency in the $X$ site and correspondingly more Si in the $Z$ site;
- Phengite with Mg or $Fe^{2+}$ substituting for Al in the $Y$ site and a corresponding increase in Si in the $Z$ site.

### Occurrence and Production

Mica is widely distributed and occurs in igneous, metamorphic and sedimentary regimes. Large crystals of mica used for various applications are typically mined from granitic pegmatites.

Until the 19th century, large crystals of mica were quite rare and expensive as a result of the limited supply in Europe. However, their price dramatically dropped when large reserves were found and mined in Africa and South America during the early 19th century. The largest documented single crystal of mica (phlogopite) was found in Lacey mine, Ontario, Canada; it measured 10×4.3×4.3 m and weighed about 330 tonnes. Similar-sized crystals were also found in Karelia, Russia.

The British Geological Survey reported that as of 2005, Koderma district in Jharkhand state in India had the largest deposits of mica in the world. China was the top producer of mica with almost a third of the global share, closely followed by the US, South Korea and Canada. Large deposits of sheet mica were mined in New England from the 19th century to the 1970s. Large mines existed in Connecticut, New Hampshire, and Maine.

Scrap and flake mica is produced all over the world. In 2010, the major producers were Russia (100,000 tonnes), Finland (68,000 t), United States (53,000 t), South Korea (50,000 t), France (20,000 t) and Canada (15,000 t). The total production was 350,000 t, although no reliable data were available for China. Most sheet mica was produced in India (3,500 t) and Russia (1,500 t). Flake mica comes from several sources: the metamorphic rock called schist as a byproduct of processing feldspar and kaolin resources, from placer deposits, and from pegmatites. Sheet mica is considerably less abundant than flake and scrap mica, and is occasionally recovered from mining scrap and flake mica. The most important sources of sheet mica are pegmatite deposits. Sheet mica prices vary with grade and can range from less than $1 per kilogram for low-quality mica to more than $2,000 per kilogram for the highest quality.

## Properties and Uses

The mica group represents 37 phyllosilicate minerals that have a layered or platy texture. The commercially important micas are muscovite and phlogopite, which are used in a variety of applications. Mica's value is based on several of its unique physical properties. The crystalline structure of mica forms layers that can be split or delaminated into thin sheets usually causing foliation in rocks. These sheets are chemically inert, dielectric, elastic, flexible, hydrophilic, insulating, lightweight, platy, reflective, refractive, resilient, and range in opacity from transparent to opaque. Mica is stable when exposed to electricity, light, moisture, and extreme temperatures. It has superior electrical properties as an insulator and as a dielectric, and can support an electrostatic field while dissipating minimal energy in the form of heat; it can be split very thin (0.025 to 0.125 millimetres or thinner) while maintaining its electrical properties, has a high dielectric breakdown, is thermally stable to 500 °C, and is resistant to corona discharge. Muscovite, the principal mica used by the electrical industry, is used in capacitors that are ideal for high frequency and radio frequency. Phlogopite mica remains stable at higher temperatures (to 900 °C) and is used in applications in which a combination of high-heat stability and electrical properties is required. Muscovite and phlogopite are used in sheet and ground forms.

### *Ground Mica*

The leading use of dry-ground mica in the US is in joint compound for filling and finishing seams and blemishes in gypsum wallboard (drywall). The mica acts as a filler and extender, provides a smooth consistency, improves the workability of the compound, and provides resistance to cracking. In 2008, joint compound accounted for 54% of dry-ground mica consumption. In the paint industry, ground mica is used as a pigment extender that also facilitates suspension, reduces chalking, prevents shrinking and shearing of the paint film, increases resistance of the paint film to water penetration and weathering, and brightens the tone of coloured pigments. Mica also promotes paint adhesion in aqueous and oleoresinous formulations. Consumption of dry-ground mica in paint, the second ranked use, accounted for 22% of the dry-ground mica used in 2008.

Ground mica is used in the well-drilling industry as an additive to drilling fluids. The coarsely ground mica flakes help prevent the loss of circulation by sealing porous sections of the drill hole. Well drilling muds accounted for 15% of dry-ground mica use in 2008. The plastics industry used dry-ground mica as an extender and filler, especially in parts for automobiles as lightweight insulation to suppress sound and vibration. Mica is used in plastic automobile fascia and fenders as a reinforcing material, providing improved mechanical properties and increased dimensional stability, stiffness, and strength. Mica-reinforced plastics also have high-heat dimensional stability, reduced warpage, and the best surface properties of any filled plastic composite. In 2008, consumption of dry-ground mica in plastic applications accounted for 2% of the market. The rubber industry used ground mica as an inert filler and mold release compound in the manufacture of molded rubber products, such as tires and roofing. The platy texture acts as an antiblocking, antisticking agent. Rubber mold lubricant accounted for 1.5% of the dry-ground mica used in 2008. As a rubber additive, mica reduces gas permeation and improves resiliency.

Dry-ground mica is used in the production of rolled roofing and asphalt shingles, where it serves as a surface coating to prevent sticking of adjacent surfaces. The coating is not absorbed by freshly manufactured roofing because mica's platy structure is unaffected by the acid in asphalt or by weather conditions. Mica is used in decorative coatings on wallpaper, concrete, stucco, and tile surfaces. It also is used as an ingredient in flux coatings on welding rods, in some special greases, and as coatings for core and mold release compounds, facing agents, and mold washes in foundry applications. Dry-ground phlogopite mica is used in automotive brake linings and clutch plates to reduce noise and vibration (asbestos substitute); as sound-absorbing insulation for coatings and polymer systems; in reinforcing additives for polymers to increase strength and stiffness and to improve stability to heat, chemicals, and ultraviolet (UV) radiation; in heat shields and temperature insulation; in industrial coating additive to decrease the permeability of moisture and hydrocarbons; and in polar polymer formulations to increase the strength of epoxies, nylons, and polyesters.

Wet-ground mica, which retains the brilliancy of its cleavage faces, is used primarily in pearlescent paints by the automotive industry. Many metallic-looking pigments are composed of a substrate of mica coated with another mineral, usually titanium dioxide ($TiO_2$). The resultant pigment produces a reflective colour depending on the thickness of the coating. These products are used to produce automobile paint, shimmery plastic containers, high quality inks used in advertising and security applications. In the cosmetics industry, its reflective and refractive properties make mica an important ingredient in blushes, eye liner, eye shadow, foundation, hair and body glitter, lipstick, lip gloss, mascara, moisturizing lotions, and nail polish. Some brands of toothpaste include powdered white mica. This acts as a mild abrasive to aid polishing of the tooth surface, and also adds a cosmetically pleasing, glittery shimmer to the paste. Mica is added to latex balloons to provide a coloured shiny surface.

Mica is also used as an insulator in concrete block, home attics, and can be poured into walls (usually in retrofitting uninsulated open top walls). Mica may also be used as a soil conditioner, especially in potting soil mixes and in gardening plots. Greases used for axles are composed of a compound of fatty oils to which mica, tar or graphite is added to increase the durability of the grease and give it a better surface.

### *Built-up Mica*

Muscovite and phlogopite splittings can be fabricated into various built-up mica products. Produced by mechanized or hand setting of overlapping splittings and alternate layers of binders and splittings, built-up mica is used primarily as an electrical insulation material. Mica insulation is used in high-temperature and fire-resistant power cables in aluminium plants, blast furnaces, critical wiring circuits (for example, defense systems, fire and security alarm systems, and surveillance systems), heaters and boilers, lumber kilns, metal smelters, and tanks and furnace wiring. Specific high-temperature mica-insulated wire and cable is rated to work for up to 15 minutes in molten aluminium, glass, and steel. Major products are bonding materials; flexible, heater, molding, and segment plates; mica paper; and tape.

Flexible plate is used in electric motor and generator armatures, field coil insulation, and magnet and commutator commutator core insulation. Mica consumption in flexible plate was about 21 tonnes in 2008 in the US. Heater plate is used where high-temperature insulation is required. Molding plate is sheet mica from which V-rings are cut and stamped for use in insulating the copper segments from the steel shaft ends of a commutator. Molding plate is also fabricated into tubes and rings for insulation in armatures, motor starters, and transformers. Segment plate acts as insulation between the copper commutator segments of direct-current universal motors and generators. Phlogopite built-up mica is preferred because it wears at the same rate as the copper segments. Although muscovite has a greater resistance to wear, it causes uneven ridges that may interfere with the operation of a motor or generator. Consumption of segment plate was about 149 t in 2008 in the US. Some types of built-up mica have the bonded splittings reinforced with cloth, glass, linen, muslin, plastic, silk, or special paper. These products are very flexible and are produced in wide, continuous sheets that are either shipped, rolled, or cut into ribbons or tapes, or trimmed to specified dimensions. Built-up mica products may also be corrugated or reinforced by multiple layering. In 2008, about 351 t of built-up mica was consumed in the US, mostly for molding plates (19%) and segment plates (42%).

### *Sheet Mica*

Sheet mica is used in electrical components, electronics, isinglass, and atomic force microscopy. Other uses include diaphragms for oxygen-breathing equipment, marker dials for navigation compasses, optical filters, pyrometres, thermal regulators, stove and kerosene heater windows, radiation aperture covers for microwave ovens, and micathermic heater elements. Mica is birefringent and is therefore commonly used to make quarter and half wave plates. Specialized applications for sheet mica are found in aerospace components in air-, ground-, and sea-launched missile systems, laser devices, medical electronics and radar systems. Mica is mechanically stable in micrometre-thin sheets which are relatively transparent to radiation (such as alpha particles) while being impervious to most gases. It is therefore used as a window on radiation detectors such as Geiger-Müller tubes.

In 2008, mica splittings represented the largest part of the sheet mica industry in the United States. Consumption of muscovite and phlogopite splittings was about 308 t in 2008. Muscovite splittings from India accounted for essentially all domestic consumption. The remainder was primarily imported from Madagascar.

Small squared pieces of sheet mica are also used in the traditional Japanese Kodo ceremony to burn incense. The sheet of mica is placed on top of a cone made of white ash, which contains a burning piece of coal, acting as a separator between the heat source and the incense, in order to spread the fragrance without burning it.

## Electrical and Electronic

Sheet mica is used principally in the electronic and electrical industries. Its usefulness in these applications is derived from its unique electrical and thermal insulating properties and its mechanical properties, which allow it to be cut, punched, stamped, and machined to close tolerances. Specifically, mica is unusual in that it is a good electrical insulator at the same time as being a good thermal conductor. The leading use of block mica is as an electrical insulator in electronic equipment. High-quality block mica is processed to line the gauge glasses of high-pressure steam boilers because of its flexibility, transparency, and resistance to heat and chemical attack. Only high-quality muscovite film mica, which is variously called India ruby mica or ruby muscovite mica, is used as a dielectric in capacitors. The highest quality mica film is used to manufacture capacitors for calibration standards. The next lower grade is used in transmitting capacitors. Receiving capacitors use a slightly lower grade of high-quality muscovite.

Mica sheets are used to provide structure for heating wire (such as in Kanthal or Nichrome) in heating elements and can withstand up to 900 °C (1,650 °F).

### *Isinglass*

Thin transparent sheets of mica called "isinglass" were used for peepholes in boilers, lanterns, stoves, and kerosene heaters because they were less likely to shatter than glass when exposed to extreme temperature gradients. Such peepholes were also used in "isinglass curtains" in horse-drawn carriages and early 20th century cars.

## Atomic Force Microscopy

Another use of mica is in the production of ultraflat, thin-film surfaces (e.g. gold surfaces) using mica as substrate. Although the deposited film surface is still rough due to deposition kinetics, the back side of the film at mica-film interface provides ultraflatness, when the film is removed from the substrate. Such ultraflat substrates are common substrates for sample preparation for the atomic force microscopy. Freshly cleaved mica surfaces have been used as clean imaging substrates in atomic force microscopy, enabling for example the imaging of bismuth films, plasma glycoproteins, membrane bilayers, and DNA molecules.

### *Substitutes*

Some lightweight aggregates, such as diatomite, perlite, and vermiculite, may be substituted for ground mica when used as filler. Ground synthetic fluorophlogopite, a fluorine-rich mica, may replace natural ground mica for uses that require thermal and electrical properties of mica. Many materials can be substituted for mica in numerous electrical, electronic, and insulation uses. Substitutes include acrylate polymers, cellulose acetate, fibreglass, fishpaper, nylon, phenolics, polycarbonate, polyester, styrene, vinyl-PVC, and vulcanized fibre. Mica paper made from scrap mica can be substituted for sheet mica in electrical and insulation applications.

### *Mica in Ancient Times*

Human use of mica dates back to prehistoric times. Mica was known to ancient Indian, Egyptian, Greek and Roman and Chinese civilizations, as well as the Aztec civilization of the New World.

The earliest use of mica has been found in cave paintings created during the Upper Paleolithic period (40,000 BC to 10,000 BC). The first hues were red (iron oxide, hematite, or red ochre) and black (manganese dioxide, pyrolusite), though black from juniper or pine carbons has also been discovered. White from kaolin or mica was used occasionally.

A few kilometres northeast of Mexico City stands the ancient site of Teotihuacan. The most striking visual and striking structure of Teotihuacan is the towering Pyramid of the Sun. The pyramid contained considerable amounts of mica in layers up to 30 cm (12 in) thick.

Natural mica was and is still used by the Taos and Picuris Pueblos Indians in north-central New Mexico to make pottery. The pottery is made from weathered Precambrian mica schist, and has flecks of mica throughout the vessels. Tewa Pueblo pottery is made by coating the clay with mica to provide a dense-glittery micaceous finish over the entire object.

Mica flakes (called Abrak in Urdu and written as ÇÈÑ˜) are also used in Pakistan to embellish women's summer clothes, especially dupattas ( light long scarves, often colourful and matching the dress). Thin mica flakes are added to hot starch water solution and the dupatta is dipped in this water mixture for 3-5 minutes. Then it is hanged to air dry.

### Mica Powder

Throughout the ages, fine powders of mica have been used for various purposes, including decorations. Powdered mica glitter is used to decorate traditional water clay pots in India, Pakistan and Bangladesh; it is also used on traditional Pueblo pottery, though not restricted to use on water pots in this case. The *gulal* and *abir* (coloured powders) used by North Indian Hindus during the festive season of Holi contain fine, small crystals of mica to create a sparkling effect. The majestic Padmanabhapuram Palace, 65 km (40 mi) from Trivandrum in India, has coloured mica windows.

## Chlorite Group

The chlorites are a group of phyllosilicate minerals. Chlorites can be described by the following four endmembers based on their chemistry via substitution of the following four elements in the silicate lattice; Mg, Fe, Ni, and Mn.

- Clinochlore: $(Mg_5Al)(AlSi_3)O_{10}(OH)_8$
- Chamosite: $(Fe_5Al)(AlSi_3)O_{10}(OH)_8$
- Nimite: $(Ni_5Al)(AlSi_3)O_{10}(OH)_8$
- Pennantite: $(Mn,Al)_6(Si,Al)_4O_{10}(OH)_8$

In addition, zinc, lithium, and calcium species are known. The great range in composition results in considerable variation in physical, optical, and X-ray properties. Similarly, the range of chemical composition allows chlorite group minerals to exist over a wide range of temperature and pressure conditions. For this reason chlorite minerals are ubiquitous minerals within low and medium temperature metamorphic rocks, some igneous rocks, hydrothermal rocks and deeply buried sediments.

### *Chlorite Structure*

The typical general formula is: $(Mg,Fe)_3(Si,Al)_4O_{10}(OH)_2 \cdot (Mg,Fe)_3(OH)_6$. This formula emphasizes the structure of the group.

Chlorites have a 2:1 sandwich structure (2:1 sandwich layer = tetrahedral-octahedral-tetrahedral = t-o-t...), this is often referred to as a talc layer. Unlike other 2:1 clay minerals, a chlorite's interlayer space (the space between each 2:1 sandwich filled by a cation) is composed of $(Mg^{2+}, Fe^{3+})(OH)_6$. This $(Mg^{2+}, Fe^{3+})(OH)_6$ unit is more commonly referred to as the brucite-like layer, due to its closer resemblance to the mineral brucite ($Mg(OH)_2$). Therefore, chlorite's structure appears as follows:

-t-o-t-brucite-t-o-t-brucite ...

An older classification divided the chlorites into two subgroups: the orthochlorites and leptochlorites. The terms are seldom used and the *ortho* prefix is somewhat misleading as the chlorite crystal system is monoclinic and not orthorhombic.

### *Occurrence*

***Figure:*** *Quartz crystal with chlorite inclusions from Minas Gerais, Brazil (size: 4.2 × 3.9 × 3.3 cm)*

Chlorite is commonly found in igneous rocks as an alteration product of mafic minerals such as pyroxene, amphibole, and biotite. In this environment chlorite may be a retrograde metamorphic alteration mineral of existing ferromagnesian minerals, or it may be present as a metasomatism product via addition of Fe, Mg, or other compounds into the rock mass. Chlorite is a common mineral associated with hydrothermal ore deposits and commonly occurs with epidote, sericite, adularia and sulfide minerals. Chlorite is also a common metamorphic mineral, usually indicative of low-grade metamorphism. It is the diagnostic species of the zeolite facies and of lower greenschist facies. It occurs in the quartz, albite, sericite, chlorite, garnet assemblage of pelitic schist. Within ultramafic rocks, metamorphism can also produce predominantly clinochlore chlorite in association with talc.

***Figure:*** *Chlorite pseudomorph after garnet from Michigan (size: 3.5 × 3.1 × 2.7 cm)*

Experiments indicate that chlorite can be stable in peridotite of the Earth's mantle above the ocean lithosphere carried down by subduction, and chlorite may even be present in the mantle volume from which island arc magmas are generated.

Chlorite occurs naturally in a variety of locations and forms. For example, chlorite is found naturally in certain parts of Wales in mineral schists. Chlorite is found in large boulders scattered on the ground surface on Ring Mountain in Marin County, California.

### *Members of the Chlorite Group*

| | |
|---|---|
| Baileychlore | $(Zn,Fe^{+2},Al,Mg)_6(Al,Si)_4O_{10}(O,OH)_8$ |
| Chamosite | $(Fe,Mg)_5Al(Si_3Al)O_{10}(OH)_8$ |
| Clinochlore | $(Mg,Fe^{2+})_5Al(Si_3Al)O_{10}(OH)_8$ |
| Cookeite | $LiAl_4(Si_3Al)O_{10}(OH)_8$ |
| Donbassite | $Al_2[Al_{2.33}][Si_3AlO_{10}](OH)_8$ |
| Gonyerite | $(Mn,Mg)_5(Fe^{+3})_2Si_3O_{10}(OH)_8$ |
| Nimite | $(Ni,Mg,Al)_6(Si,Al)_4O_{10}(OH)_8$ |
| Odinite | $(Fe,Mg,Al,Fe,Ti,Mn)_{2.4}(Al,Si)_2O_5OH_4$ |
| Orthochamosite | $(Fe^{+2},Mg,Fe^{+3})_5Al(Si_3Al)O_{10}(O,OH)_8$ |
| Pennantite | $(Mn_5Al)(Si_3Al)O_{10}(OH)_8$ |
| Ripidolite | $(Mg,Fe,Al)_6(Al,Si)_4O_{10}(OH)_8$ |
| Sudoite | $Mg_2(Al,Fe)_3Si_3AlO_{10}(OH)_8$ |

Clinoclore, pennantite, and chamosite are the most common varieties. Several other sub-varieties have been described. A massive compact variety of clinochlore used as a decorative carving stone is referred to by the trade name seraphinite. It occurs in the Korshunovskoye iron skarn deposit in the Irkutskaya Oblast of Eastern Siberia.

The name *chlorite* is from the Greek *chloros* meaning "green", in reference to its colour.

### *Distinguishing from Other Minerals*

Chlorite is so soft that it can be scratched by a finger nail. The powder generated by scratching is green. It feels oily when rubbed between the fingers. The plates are flexible, but not elastic like mica.

Talc is much softer and feels soapy between fingers. The powder generated by scratching is white.

Mica plates are elastic whereas chlorite plates are flexible without bending back.

# Bibliography


Aitchison, Leslie: *A History of Metals*: Interscience Publishers, New York, 1960.

Bates, Robert Latimer: *Geology of the Industrial Rocks and Minerals*, Dover Publications, New York, 1969.

Belknap, D.F.: *Sea-level Rise on the Maine Islands: Proceedings*, Sixth Island Institute Conference, Hurricane Island, Maine, Island Institute, Rockland, ME, 1988.

Binay Kumar and R.P. Tandon: *Advances in Technologically Important Crystals*, Macmillan Publishers India, Delhi, 2007.

Brookins, Douglas G.: *Mineral and Energy Resources: Occurrence, Exploitation, and Environmental Impact*, Merrill Pub. Co., Columbus, 1990.

Cotton, C.A.: *Volcanoes as Landscape Forms*, Hafner Publishing. New York, 1999.

Donald D. Carr.: *Industrial Minerals and Rocks,* Society for Mining, Metallurgy, and Exploration, Littleton, Colo, 1994.

Doyle, F. L.: *Karst Hydrogeology,* The University of Alabama in Huntsville Press, Huntsville, Alabama, 1977.

Edwards, Denis: *Earth Revealing Earth Healing: Ecology and Christian Theology,* Collegeville: Liturgical Press, 2001.

Edwards, Edward: *Pattern and Design with Dynamic Symmetry,* NY: Dover 1967.

Gontz, A.M.: *Evolution of Seabed Pockmarks in Penobscot Bay, Maine*, University of Maine, Orono, 2011.

Gurjar, Ram Kumar: *Geography of Water Resources*, Rawat, Delhi, 2008.

Hartman, Howard L.: *Introductory Mining Engineering*, Wiley, New York, c1987.

Holmgren, D. A., Moody, J. D.: *The Structural Setting for Giant oil and Gas Fields*, Tokyo, 1975.

John C. Grover.: *Volcanic Eruptions and Great Earthquakes*, Copyright Publishing Company, Australia, 1998.

Kesler, Stephen E.: *Mineral Resources, Economics, and the Environment*, Maxwell Macmillan International, New York, 1994.

Lovesun, V.J. ; P.K. Sen and Amalendu Sinha: *Exploration, Exploitation, Enrichment and Environment of Coastal Placer Minerals*, Macmillan Publishers India, Delhi, 2007.

Majumdar P.K. Basu: *Law of Mines and Minerals*, Universal Law, Delhi, 2011.

Mary Lambert: *Crystal Energy*, Cico Books, Delhi, 2010.

Peters, William C.: *Exploration and Mining Geology*, J. Wiley, New York, 1987.

Poonam Johri: *Minerals and Water*, Sonali, Delhi, 2005.

Pramod O. Alexander: *A Handbook of Minerals, Crystals, Rocks and Ores*, New India Pub Agency, Delhi, 2009.

Prieto, Carlos: *Mining in the New World*, McGraw-Hill, New York, 1973.

Roberts, Willard Lincoln, Thomas J. Campbell, and George Robert Rapp.: *Encyclopedia of Minerals,* Van Nostrand Reinhold, New York, 1990.

Sawant, P. T.: *Engineering and General Geology*, New India Publishing Agency, Delhi, 2011.

Scalisi, Philip, and David Cook: *Classic Mineral Localities of the World, Asia and Australia,* Van Nostrand Reinhold Co., New York, 1983.

Shepherd, R.: *Ancient Mining*, Elsevier Applied Science, London, New York, c1993.

Tiwari, S.K.: *Ore Geology, Economic Minerals And Mineral Economics*, Atlantic Publishers, Delhi, 2010.

Tracy, Robert J.; Owens, Brent: *Petrology: Igneous, Sedimentary, and Metamorphic,* New York: W. H. Freeman, 2005.

Vaughan, D. J.; Craig, J. R.: *Mineral Chemistry of Metal Sulfides*, Cambridge University Press, Cambridge, 1978.

Wahab, M. A.: *Essentials of Crystallography*, Narosa, Delhi, 2009.

Whitmore, G: *Chemical Crystallography And Lipid Crystals*, Sarup Book, Delhi, 2009.

Wolfe, John A: *Mineral Resources: A World Review*, Chapman & Hall, New York, 1984.

# Index

❑❑❑